Voyager

RAYMOND E. FEIST

Krondor:
The Betrayal

Book I of The Riftwar Legacy

Based on the game
Betrayal at Krondor
published by Dynamix, Inc.
Story by Neal Hallford, John Cutter,
and Raymond E. Feist

HarperCollins*Publishers*

Voyager
An Imprint of HarperCollins*Publishers*
77–85 Fulham Palace Road,
Hammersmith, London W6 8JB
www. voyager-books.com

Special overseas edition 1999
This paperback edition 1999
11

First published in Great Britain by
Voyager1998

ISBN 0-00-648334-8

Set in Meridien

Printed in Great Britain by
Clays Ltd, St Ives plc

Acknowledgements

Again I am in debt to many people.

The original Midkemians, for the universe in which I work, and for their understanding of what makes a good story, a good game, and how the two are different.

My agent Jonathan Matson, for shepherding me through major difficulties in creating these games, with his usual deft touch and quick wit.

John Cutter, who thought it up in the first place.

Neal Hallford, who created a very nifty story for the core of the game which provided the basis for this book.

The rest of the creative team at Dynamix who managed to squeeze the most out of the processor to give us music, pictures, sound and story.

And to Jerry Lutrell, for keeping me apprised of what was what early on.

My wife, Kathlyn S. Starbuck, for being who she is.

My children Jessica and James, for keeping me in touch with what's important daily and for being the most wonderful children any father could ask for.

Raymond E. Feist
Rancho Santa Fe, CA
March 11, 1998

For John Cutter and Neal Hallford
with thanks for their creativity and enthusiasm

CONTENTS

Warning

The wind howled.

Locklear, squire of the Prince of Krondor's court, sat huddled under his heavy cloak, astride his horse. Summer was quick to flee in the Northlands and the passes through the mountains known as the Teeth of the World. Autumn nights in the south might still be soft and warm, but up here in the north, autumn had been a brief visitor and winter was early to arrive, and would be long in residence. Locklear cursed his own stupidity for leading him to this forlorn place.

Sergeant Bales said, 'Gets nippy up here, squire.' The sergeant had heard the rumour about the young noble's sudden appearance in Tyr-Sog, some matter involving a young woman married to a well-connected merchant in Krondor. Locklear wouldn't be the first young dandy sent to the frontier to get him out of an angry husband's reach. 'Not as balmy as Krondor, sorry to say, sir.'

'Really?' asked the young squire, dryly.

The patrol followed a narrow trail along the edge of the foothills, the northern border of the Kingdom of the Isles. Locklear had been in court at Tyr-Sog less than a week when Baron Moyiet had suggested the young squire might benefit from accompanying the special patrol to the east of the city. Rumours had been circulating that renegades and moredhel – dark elves known as the Brotherhood of the Dark Path – were infiltrating south under the cover of heavy rains and snow flurries. Trackers had reported few signs, but hearsay and the insistence of farmers that they had seen companies of dark-clad warriors hurrying south had prompted the Baron to order the patrol.

Locklear knew as well as the men garrisoned there that the chance of any activity along the small passes over the mountains in late fall or early winter was unusual. While the freeze had just come to the foothills, the higher passes would already be thick with snow, then choked with mud should a brief thaw occur.

Yet since the war known as the Great Uprising – the invasion of the Kingdom by the army of Murmandamus, the charismatic leader of the dark elves – ten years ago, any activity was to be investigated, and that order came directly from King Lyam.

'Yes, must be a bit of a change from the Prince's court, squire,' prodded the sergeant. Locklear had looked the part of a Krondorian dandy – tall, slender, a finely garbed young man in his mid-twenties, affecting a moustache and long ringlets – when he reached Tyr-Sog. Locklear thought the moustache and fine clothing made him look older, but if anything the impact was the opposite of his desired intent.

Locklear had enough of the sergeant's playful baiting, and observed, 'Still, it's warmer than I remember the other side of the mountains being.'

'Other side, sir?' asked the sergeant.

'The Northlands,' said Locklear. 'Even in the spring and summer the nights are cold.'

The sergeant looked askance at the young man. 'You've been there, squire?' Few men who were not renegades or weapons runners had visited the Northlands and lived to return to the Kingdom.

'With the Prince,' replied Locklear. 'I was with him at Armengar and Highcastle.'

The sergeant fell silent and looked ahead. The soldiers nearest Locklear exchanged glances and nods. One whispered to the man behind him. No soldier living in the north hadn't heard of the fall of Armengar before the hosts of

Murmandamus, the powerful moredhel leader who had destroyed the human city in the Northlands and then had invaded the Kingdom. Only his defeat at Sethanon, ten years before, had kept his army of dark elves, trolls, goblins and giants from rending the Kingdom.

The survivors of Armengar had come to live in Yabon, not far from Tyr-Sog, and the telling of the great battle and the flight of the survivors, as well as the part played by Prince Arutha and his companions, had grown in the telling. Any man who had served with Prince Arutha and Guy du Bas-Tyra could only be judged a hero. With a reappraising glance at the young man, the sergeant kept his silence.

Locklear's amusement at shutting up the voluble sergeant was shortlived, as the snow started to freshen, blowing harder by the minute. He might have gained enough stature with the garrison to be treated with more respect in days to come, but he was still a long way from the court in Krondor, the fine wines and pretty girls. It would take a miracle for him to get back in Arutha's good graces any time before the next winter found him still trapped in a rural court with dullards.

After ten minutes of silent travel, the sergeant said, 'Another two miles, sir, and we can start back.'

Locklear said nothing. By the time they returned to the garrison, it would be dark and even colder than it presently was. He would welcome the warm fire in the soldiers' commons and probably content himself sharing a meal with the troops, unless the Baron requested he dine with the household. Locklear judged that unlikely, as the Baron had a flirtatious young daughter who had fawned on the visiting young noble the first night he had appeared in Tyr-Sog, and the Baron full well knew why Locklear was at his court. On the two occasions he had since dined with the Baron, the daughter had been conspicuously absent.

There was an inn not too far from the castle, but by the

time he had returned to the castle, he knew he would be too sick of the cold and snow to brave the elements again, even for that short distance; besides, the only two barmaids there were fat and dull. With a silent sigh of resignation, Locklear realized that by the arrival of spring they might look lovely and charming to him.

Locklear just prayed he would be permitted to return to Krondor by the Midsummer Festival of Banapis. He would write to his best friend, Squire James, and ask him to use his influence to get Arutha to recall him early. Half a year up here was punishment enough.

'Seigneur,' said Sergeant Bales, using Locklear's formal title, 'what's that?' He pointed up the rocky path. Movement among the rocks had caught the sergeant's eye.

Locklear replied, 'I don't know. Let's go take a look.'

Bales motioned and the patrol turned left, moving up the path. Quickly the scene before them resolved itself. A lone figure, on foot, hurried down the rocky path, and from behind the sounds of pursuit could be heard.

'Looks like a renegade had a falling-out with some Brothers of the Dark Path,' said Sergeant Bales.

Locklear pulled his own sword. 'Renegade or not, we can't let the dark elves carve him up. It might make them think they could come south and harass common citizens at whim.'

'Ready!' shouted the sergeant and the veteran patrol pulled swords.

The lone figure saw the soldiers, hesitated a moment, then ran forward. Locklear could see he was a tall man, covered by a dark grey cloak which effectively hid his features. Behind him on foot came a dozen dark elves.

'Let us go amongst them,' said the sergeant calmly.

Locklear commanded the patrol in theory, but he had enough combat experience to stay out of the way when a veteran sergeant was giving orders.

The horsemen charged up the pass, moving by the lone

figure, to fall upon the moredhel. The Brotherhood of the Dark Path were many things; cowardly and inept in warcraft were not among those things. The fighting was fierce, but the Kingdom soldiers had two advantages: horses, and the fact the weather had rendered the dark elves' bows useless. The moredhel didn't even attempt to draw their wet strings, knowing they could hardly send a bowshaft toward the enemy, let alone pierce armour.

A single dark elf, larger than the rest, leaped atop a rock, his gaze fixed upon the fleeing figure. Locklear moved his horse to block the creature, who turned his attention toward the young noble.

They locked gazes for a moment, and Locklear could feel the creature's hatred. Silently he seemed to mark Locklear, as if remembering him for a future confrontation. Then he shouted an order and the moredhel began their withdrawal up the pass.

Sergeant Bales knew better than to pursue into a pass when he had less than a dozen yards' visibility. Besides, the weather was worsening.

Locklear turned to find the lone figure leaning against a boulder a short distance behind the trail. Locklear moved his horse close to the man and called down, 'I am Squire Locklear of the Prince's court. You better have a good story for us, renegade.'

There was no response from the man, his features still hidden by the deep cowl of his heavy cloak. The sounds of fighting trailed off as the moredhel broke off and fled up the pass, crawling into the rocks above the path so the riders could not follow.

The figure before Locklear regarded him a moment, then slowly reached up to throw back his cowl. Dark, alien eyes regarded the young noble. These were features Locklear had seen before: high brow, close-cropped hair. Arching eyebrows and large, upswept and lobeless ears. But this was no elf who stood before him; Locklear could feel it in his bones. The dark eyes that regarded him could barely hide their contempt.

In heavily accented King's Tongue, the creature said, 'I am no renegade, human.'

Sergeant Bales rode up and said, 'Damn! A Brother of the Dark Path. Must have been some tribal thing, with those others trying to kill him.'

The moredhel fixed Locklear with his gaze, studying him for a long moment, then he said, 'If you are from the Prince's court then you may help me.'

'Help you?' said the sergeant. 'We're most likely going to hang you, murderer.'

Locklear held up his hand for silence. 'Why should we help you, moredhel?'

'Because I bring a word of warning for your prince.'

'Warning of what?'

'That is for him to know. Will you take me to him?'

Locklear glanced at the sergeant, who said, 'We should take him to see the Baron.'

'No,' said the moredhel. 'I will only speak with Prince Arutha.'

'You'll speak to whoever we tell you to, butcher!' said Bales, his voice edged in hatred. He had been fighting the Brotherhood of the Dark Path his entire life and had seen their cruelty many times.

Locklear said, 'I know his kind. You can set fire to his feet and burn him up to his neck and if he doesn't want to talk, he won't talk.'

The moredhel said, 'True.' He again studied Locklear and said, 'You have faced my people?'

'Armengar,' said Locklear. 'Again at Highcastle. Then at Sethanon.'

'It is Sethanon about which I need to speak to your prince,' said the moredhel.

Locklear turned to the sergeant and said, 'Leave us for a moment, Sergeant.'

Bales hesitated, but there was a note of command in the young noble's voice, no hint of deference to the sergeant; this was an order. The sergeant turned and moved his patrol away.

'Say on,' said Locklear.

'I am Gorath, Chieftain of the Ardanien.'

Locklear studied Gorath. By human standards he looked young, but Locklear had been around enough elves and seen enough moredhel to know that was deceiving. This one had a beard streaked with white and grey, as well as a few lines around his eyes; Locklear guessed he might be better than two hundred years old by what he had seen among elvenkind. Gorath wore armour that was well crafted and a cloak of especially fine weave; Locklear judged it possible he was exactly what he said he was. 'What does a moredhel chieftain speak of to a prince of the Kingdom?'

'My words are for Prince Arutha alone.'

Locklear said, 'If you don't want to spend what remains of your life in the Baron's dungeon at Tyr-Sog, you had better say something that will convince me to take you to Krondor.'

The moredhel looked a long time at Locklear, then motioned for him to come closer. Keeping his hand upon a dagger in his belt, should the dark elf try something, he leaned close to his horse's neck, so he could put his face near Gorath's.

Gorath whispered in Locklear's ear. 'Murmandamus lives.'

Locklear leaned back and was silent a moment, then he turned his horse. 'Sergeant Bales!'

'Sir!' returned the old veteran, answering Locklear's commanding tone of voice with a note of respect.

'Put this prisoner in chains. We return to Tyr-Sog, now. And no one is to speak with him without my leave.'

'Sir!' repeated the sergeant, motioning to two of his men to hurry forward and do as ordered.

Locklear leaned over his horse's neck again and said, 'You may be lying to stay alive, Gorath, or you may have some dreadful message for Prince Arutha. It matters not to me, for either way I return to Krondor, starting first thing in the morning.'

The dark elf said nothing, content to stand stoically as he was disarmed by two soldiers. He remained silent as manacles were fastened around his wrists, linked by a short span of heavy chain. He held his hands before him a moment after the manacles were locked, then slowly lowered them. He looked at Locklear, then turned and began walking toward Tyr-Sog, without waiting for his guards' leave.

Locklear motioned for the sergeant to follow, and rode up to walk his horse next to Gorath, through the worsening weather.

ONE

Encounter

The fire crackled.

Owyn Belefote sat alone in the night before the flames, wallowing in his personal misery. The youngest son of the Baron of Timons, he was a long way from home and wishing he was even farther away. His youthful features were set in a portrait of dejection.

The night was cold and the food scant, especially after having just left the abundance of his aunt's home in Yabon City. He had been hosted by relatives ignorant of his falling-out with his father, people who had reacquainted him over a week's visit with what he had forgotten about his home-life: the companionship of brothers and sisters, the warmth of a night spent before the fire, conversation with his mother, and even the arguments with his father.

'Father,' Owyn muttered. It had been less than two years since the young man had defied his father and made his way to Stardock, the island of magicians located in the southern reaches of the Kingdom. His father had forbidden him his choice, to study magic, demanding Owyn should at least become a cleric of one of the more socially acceptable orders of priests. After all, they did magic as well, his father had insisted.

Owyn sighed and gathered his cloak around him. He had been so certain he would someday return home to visit his family, revealing himself as a great magician, perhaps a confidant of the legendary Pug, who had created the Academy at Stardock. Instead he found himself ill-suited for the study required. He also had no love for the burgeoning politics of the place, with factions of students rallying around this teacher or that, attempting to turn the study of magic into another religion. He now knew he was, at best, a mediocre magician and would never amount to more, and no matter how much he wished to study magic, he lacked sufficient talent.

After slightly more than one year of study, Owyn had left Stardock, conceding to himself that he had made a mistake. Admitting such to his father would prove a far more daunting task – which was why he had decided to visit family in the distant province of Yabon before mustering the courage to return to the east and confront his sire.

A rustle in the bushes caused Owyn to clutch a heavy wooden staff and jump to his feet. He had little skill with weapons, having neglected that portion of his education as a child, but had developed enough skill with this quarterstaff to defend himself.

'Who's there?' he demanded.

From out of the gloom came a voice, saying, 'Hello, the camp. We're coming in.'

Owyn relaxed slightly, as bandits would be unlikely to warn him they were coming. Also, he was obviously not worth attacking, as he looked little more than a ragged beggar these days. Still, it never hurt to be wary.

Two figures appeared out of the gloom, one roughly Owyn's height, the other a head taller. Both were covered in heavy cloaks, the smaller of the two limping obviously.

The limping man looked over his shoulder, as if being followed, then asked, 'Who are you?'

Owyn said, 'Me? Who are you?'

The smaller man pulled back his hood and said, 'Locklear, I'm a squire to Prince Arutha.'

Owyn nodded. 'Sir, I'm Owyn, son of Baron Belefote.'

'From Timons, yes, I know who your father is,' said Locklear. Squatting before the fire, opening his hands to warm them. He glanced up at Owyn. 'You're a long way from home, aren't you?'

'I was visiting my aunt in Yabon,' said the blond youth. 'I'm now on my way home.'

'Long journey,' said the muffled figure.

'I'll work my way down to Krondor, then see if I can travel with a caravan or someone else to Salador. From there I'll catch a boat to Timons.'

'Well, we could do worse than stick together until we reach LaMut,' said Locklear, sitting down heavily on the ground. His cloak fell open and Owyn saw blood on the young man's clothing.

'You're hurt,' he said.

'Just a bit,' admitted Locklear.

'What happened?'

'We were jumped a few miles north of here,' said Locklear.

Owyn started rummaging through his travel bag. 'I have something in here for wounds,' he said. 'Strip off your tunic.'

Locklear removed his cloak and tunic while Owyn took bandages and powder from his bag. 'My aunt insisted I take this just in case. I thought it an old lady's foolishness, but apparently it wasn't.'

Locklear endured the boy's ministrations as he washed the wound – obviously a sword cut to the ribs – and winced when the powder was sprinkled upon it. Then as he bandaged the squire's ribs, Owyn said, 'Your friend doesn't talk much, does he?'

'I am not his friend,' answered Gorath. He held out his manacles for inspection. 'I am his prisoner.'

Trying to peer into the darkness of Gorath's hood, Owyn said, 'What did he do?'

'Nothing, except be born on the wrong side of the mountains,' offered Locklear.

Gorath pulled back his hood, and graced Owyn with the faintest of smiles.

'Gods' teeth!' exclaimed Owyn. 'He's a Brother of the Dark Path!'

'Moredhel,' corrected Gorath, with a note of ironic bitterness. '"Dark elf", in your tongue, human. At least our cousins in Elvandar would have you believe us so.'

Locklear winced as Owyn applied his aunt's salve to the wounded ribs. 'A couple of hundred years of war lets us form our own opinions, thank you, Gorath.'

Gorath said, 'You understand so little, you humans.'

'Well,' said Locklear, 'I'm not going anywhere at the moment, so educate me.'

Gorath looked at the young squire, as if trying to judge something, and was silent for a while. 'Those you call "elves" and my people are one, by blood, but we live different lives. We were the first mortal race after the great dragons and the Ancient Ones.'

Owyn looked at Gorath in curiosity, while Locklear just gritted his teeth and said, 'Hurry it up, would you, lad?'

'Who are the Ancient Ones?' asked Owyn in a whisper.

'The Dragon Lords,' said Locklear.

'Lords of power, the Valheru,' supplied Gorath. 'When they departed this world, they placed our fate in our own hands, naming us a free people.'

Locklear said, 'I've heard the story.'

'It is more than a story, human, for to my people it gave over this world to our keeping. Then came you humans, and the dwarves, and others. This is our world and you seized it from us.'

Locklear said, 'Well, I'm not a student of theology, and my knowledge of history is sadly lacking, but it seems to me that whatever the cause of our arrival on this world according to your lore, we're here and we don't have anywhere else to go. So if your kin, the elves, can make the best of it, why can't you?'

Gorath studied the young man, but said nothing. Then he stood, moving with deadly purpose toward Locklear.

Owyn had just tied off the bandage and fell hard as Locklear pushed him aside while he attempted to rise and draw his sword as Gorath closed on him.

But rather than attack Locklear, he lunged past the pair of humans, lashing out above Locklear's head with the chain that held his manacles. A ringing of steel caused Locklear to flinch aside as Gorath shouted, 'Assassin in the camp!' Then Gorath kicked hard at Owyn, shouting, 'Get out from underfoot!'

Owyn didn't know where the assassin came from; one moment there had been three of them in the small clearing, then the next Gorath was locked in a life-and-death struggle with another of his kind.

Two figures grappled by the light of the campfire, their features set in stark relief by the firelight and darkness of the woods. Gorath had knocked the other moredhel's sword from his hand, and when the second dark elf attempted to pull a dagger, Gorath slipped behind him, wrapping his wrist chains around the attacker's throat. He yanked hard and the attacker's eyes bulged in shock as Gorath said, 'Do not struggle so, Haseth. For old times' sake I will make this quick.' With a snap of his wrists, he crushed the other dark elf's windpipe, and the creature went limp.

Gorath let him fall to the ground, saying, 'May the Goddess of Darkness show you mercy.'

Locklear stood up. 'I thought we had lost them.'

'I knew we had not,' said Gorath.

'Why didn't you say something?' demanded Locklear as he retrieved his tunic and put it on over the new bandages.

'We had to turn and face him some time,' said Gorath, resuming his place. 'We could do it now, or in a day or two when you were even weaker from loss of blood and no food.' Gorath looked into the darkness from which the assassin had come. 'Had he not been alone, you'd have had only my body to drag before your prince.'

'You don't get off that easily, moredhel. You don't have

my permission to die yet, after the trouble I've gone through to keep you alive so far,' said Locklear. 'Is he the last?'

'Almost certainly not,' said the dark elf. 'But he is the last of this company. Others will come.' He glanced in the opposite direction. 'And others may already be ahead of us.'

Locklear reached into a small pouch at his side and produced a key. 'Then I think you'd better get those chains off,' he said. He unlocked the wrist irons and Gorath watched them fall to the ground with an impassive expression. 'Take the assassin's sword.'

'Maybe we should bury him?' suggested Owyn.

Gorath shook his head. 'That is not our way. His body is but a shell. Let it feed the scavengers, return to the soil, nourish the plants, and renew the world. His spirit has begun its journey through darkness, and with the Goddess of Darkness's pleasure, he may find his way to the Blessed Isles.' Gorath looked northward, as if seeking sight of something in the dark. 'He was my kinsman, though one of whom I was not overly fond. But ties of blood run strong with my people. For him to hunt me names me outcast and traitor to my race.' He looked at Locklear. 'We have common cause, then, human. For if I am to carry out the mission that brands me anathema to my people, I must survive. We need to help one another.' Gorath took Haseth's sword. To Owyn he said, 'Don't bury him, but you could pull him out of the way, human. By morning he's going to become even more unpleasant to have nearby.'

Owyn looked uncertain about touching a corpse, but said nothing as he went over, reached down and gripped the dead moredhel by the wrists. The creature was surprisingly heavy. As Owyn started to drag Haseth away, Gorath said, 'And see if he dropped his travel bag back there in the woods before he attacked us, boy. He may have something to eat in it.'

Owyn nodded, wondering what strange chance had

brought him to dragging a corpse through the dark woods and looting the body.

Morning found a tired trio making their way through the woodlands, staying within sight of the road, but not chancing walking openly along it.

'I don't see why we didn't return to Yabon and get some horses,' complained Owyn.

Locklear said, 'We have been jumped three times since leaving Tyr-Sog. If others are coming after us, I'd rather not walk right into them. Besides, we may find a village between here and LaMut where we can get some horses.'

'And pay for them with what?' asked Owyn. 'You said the fight where you were wounded was when your horses ran off with all your things. I assume that means your funds, too? I certainly don't have enough to buy three mounts.'

Locklear smiled. 'I'm not without resources.'

'We could just take them,' offered Gorath.

'There is that,' agreed Locklear. 'But without obvious badges of rank or a patent from the Prince on my person, it might prove difficult to convince the local constable of my bona fides. And we should hardly be safe penned up in a rural jail with cutthroats out looking for us.'

Owyn fell silent. They had been walking since sun-up and he was tired. 'How about a rest?' he offered.

'I don't think so,' said Gorath, his voice falling to a whisper. 'Listen.'

Neither human said anything for a moment, then Owyn said, 'What? I don't hear anything.'

'That's the point,' said Gorath. 'The birds in the trees ahead suddenly stopped their songs.'

'A trap?' asked Locklear.

'Almost certainly,' said Gorath, pulling the sword he had taken from his dead kinsman.

Locklear said, 'My side burns, but I can fight.' To Owyn he said, 'What about you?'

Owyn hefted his wooden staff. It was hard oak, with iron-shod ends. 'I can swing this, if I need to. And I have some magic.'

'Can you make them vanish?'

'No,' said Owyn. 'I can't do that.'

'Pity,' said Locklear. 'Then try to stay out of the way.'

They advanced cautiously, and as they neared the spot Gorath had indicated, Locklear could make out a shadowy figure between the trees. The man or moredhel – Locklear couldn't tell which – moved slightly, revealing his position. Had he remained motionless, Locklear would never have seen him.

Gorath signalled for Locklear and Owyn to move more to their right, looping around behind the lookout. Without knowing how many men they faced, they would do well to seek the advantage of surprise.

Gorath moved through the woods like a spirit, silent and almost unseen once Owyn and Locklear left him. Locklear signalled for Owyn to keep slightly behind and to the right of him, so he knew where he was when they closed upon their ambushers.

As they moved through the woods, they heard the sound of whispers, and Locklear knew no elves waiting for them would utter a word. Now the question was were these mere bandits or agents seeking to stop Gorath's journey.

A grunt from ahead signalled Gorath's first contact with the ambushers. A shout followed instantly and Locklear and Owyn ran forward.

Four men stood and one was already dying. The other three spread out in a small clearing between two lines of trees, a perfect position for a roadside ambush. Locklear felt an odd flicker behind him and something sped past his eyes, as if an arrow had been fired from behind, but other than the sensation of motion, there was nothing to be seen.

One of the three remaining ambushers cried out in shock,

his hand going out before him as vacant eyes stared ahead, 'I'm blind!' he shouted in panic.

Locklear decided it was Owyn's useful magic, and thanked the Goddess of Luck the boy had that much talent.

Gorath was engaged with one man while Locklear advanced on the other. Suddenly their garb registered and he said, 'Quegans!'

The men were wearing short tunics and leggings, and cross-gartered sandals. The man facing Locklear had his head covered with a red bandanna, and over his shoulder hung a baldric from which a cutlass had hung. The cutlass was now carving through the air at Locklear's head.

He parried and the blow shot fire through his wounded side. Putting aside his pain, Locklear riposted and the pirate fell back. A strangled cry told Locklear the second pirate was down.

The strange missile sensation sped by and the man facing Locklear winced and held his hand up as if shielding his eyes. Locklear didn't hesitate and ran the man through.

Gorath killed the last man and suddenly it was quiet again in the woods.

Locklear's side was afire but he didn't feel any additional damage. He put up his sword and said, 'Damn me.'

'Are you hurt?' asked Owyn.

'No,' answered Locklear.

'Then what is the problem?' asked Owyn.

Locklear looked around the clearing. '*These* are the problem. Someone has gotten word ahead of us. We can be certain of that.'

'How?' asked Gorath.

'These are Quegan pirates,' said Locklear. 'Look at their weapons.'

'I wouldn't know a Quegan if I tripped over him,' said Owyn. 'I'll take your word for it, squire.'

'Do not pirates usually ply their trade at sea?' asked Gorath.

'They do,' said Locklear, 'unless someone's paid them to stake out a road and wait for three travellers on foot.' He knelt next to the man who had died at his feet and said, 'Look at his hands. Those are the hands of a man used to handling rope. Those Quegan cutlasses are the clincher.' He examined the man, looking for a pouch or purse, saying, 'Look for anything that might be a message.'

They did and came away with a little gold and a couple of daggers in addition to the four cutlasses. But no messages or notes, nothing indicating who had hired the pirates. 'We're not close enough to Ylith for a band of pirates to have made it this far north undetected in the time since we left Yabon.'

'Someone must have sent word south when I left the Northlands,' said Gorath.

'But how?' asked Owyn. 'You've told me you only spent a couple of days in Tyr-Sog, and you were riding until yesterday.'

'That's an odd question for a student of magic,' observed Gorath.

Owyn blushed a little. 'Oh.'

'You've Spellweavers who can do such?' asked Locklear.

'Not such as the eledhel – those you call "elves" – call Spellweavers. But we have our practitioners of magic. And there are others of your race who will sell their arts.'

Owyn said, 'I've never witnessed it, but I have heard of a talent called "mind speech" which allows a spell-caster to speak with another. And there's something known as "dream speech" as well. Either – '

'Someone really wants you dead, don't they?' observed Locklear, interrupting the boy.

'Delekhan,' said Gorath. 'And he was gathering to his side any of my people who showed such talents. I know his goals, but not his plan. And if magic arts are part of it, I fear the results.'

Locklear said, 'I understand that. I've had my share of encounters with people using magic who shouldn't.' He glanced at Owyn and said, 'That blinding trick was quite good, lad.'

Looking embarrassed, Owyn said, 'I thought it might help. I know a few spells like that, but nothing that would overpower an enemy. Still, I'll try to help where I can.'

Glancing at Owyn, Locklear said, 'I know. Let's get to LaMut.'

LaMut stood astride the road south, requiring anyone travelling from Yabon to Ylith to pass through its gates or endure a long trek to the east through dangerous foothills.

The foulbourgh of the city sprawled in all directions, while the old walls of the city stood behind, nearly useless now, given the ease with which any attacker could mount the buildings next to them and gain the parapet from their roofs.

It was nearly sundown and all three travellers were tired, footsore, and hungry. 'We can present ourselves to Earl Kasumi tomorrow.'

'Why not now?' asked Owyn. 'I could use a meal and a bed.'

'Because the garrison is up there,' said Locklear, pointing at a distant fortress high above the city on a hillside, 'and that would be another two hours' walk, whereas a cheap inn is but one minute that way.' He pointed at the gate.

'Will your countrymen object to my presence?' asked Gorath.

'They would if they suspected your nature. If they think you an elf from Elvandar, they may only stare a little. Come on. We've looted enough gold for a night of relative comfort, and in the morning we'll visit the Earl and see if he can get us safely to Krondor.'

They entered the city under the watchful gaze of otherwise bored-looking soldiers. One of them stood out from his companions, being shorter, and much more businesslike in his manner. Locklear smiled and nodded at the guards, but the three travellers didn't stop or speak. A short distance inside the city gates sat an inn, marked by a wagon wheel painted bright blue. 'There,' said Locklear.

They entered the inn, busy, but not crowded, and moved to a table near the far wall. As they sat a stout young serving woman came, took their order for food and ale, and left. As they were waiting, Locklear spied a figure on the other side of the room staring at him.

It took a moment for Locklear to realize the figure wasn't a man, but a dwarf. The dwarf stood and made his way across the room. He bore a large scar across his face, cutting through his left eye. He stood before them and said, 'You don't recognize me, do you, Locky?'

Locklear realized the last time he had seen the dwarf he had not borne the scar he now sported, but at hearing his name from the dwarf's lips, he said, 'Dubal! Without the eye-patch, it took me a moment.'

The dwarf moved to sit next to Owyn, across from Gorath. 'I won this face in battle, from one of his kin – ' he pointed at Gorath ' – and I'll be a dragon's mother before I hide it again.'

'Dubal found me hiding in a cellar after the Battle of Sethanon,' said Locklear.

'Locked in there with a pretty wench, if memory serves.' The dwarf laughed.

Locklear shrugged. 'Well, that was by chance.'

Dubal said, 'Now tell me, what is a seigneur of the Prince's court doing sitting in LaMut with a moredhel warchief?' He kept his voice low, but Owyn glanced around to see if anyone had overheard him.

'You know me?' asked Gorath.

'I know your race, for you are the enemy of my blood, and I know your armour for what it is. A human might not notice, but we of the Grey Towers have fought your kind long enough I wouldn't mistake you for one from Elvandar. It's only your present company that keeps me from killing you here and now.'

Locklear held up his hand. 'I would count it a kindness and a personal favour, as would Prince Arutha, should you imagine this person on my left to be an elf.'

'I think I can manage. But you'll have to come to the Grey Towers and tell me the story behind this mummery.'

'If I can, I will,' said Locklear. 'Now, what brings you alone to LaMut?'

'We've got problems at our mines and had a collapse. Some of us are stuck on this side of the Grey Towers and I came in to the city to buy some stores. I'll hire a waggon and head back in the morning. For the time being, I'm content to sit and drink, and jabber with some of these Tsurani here in LaMut. I fought them during the war, and they've turned out to be a stalwart enough bunch once you get to know them.' He pointed to the bar. 'That tall fellow – ' Locky laughed to hear anyone call a Tsurani 'tall' ' – he's Sumani, the owner. Has a fair number of tales to spin about his days serving on the Tsurani world, and I'm switched if it doesn't sound like he's telling the truth most of the time.'

Locklear laughed. 'Most Tsurani I know don't indulge in tall tales, Dubal.'

'Seems to be so, but you never know. I've fought the big bugs, the Cho-ja, but some of those other things he talks about, well, I'm hard-pressed to believe them.'

The serving woman arrived with the food and ale and they fell to. 'Now,' said Dubal, 'can you tell me what brings you here?'

'No,' replied Locklear, 'but we can ask you if you've seen any Quegans hanging around?'

'There was a pack of them through here two days ago, according to the gossip,' said Dubal. 'I just arrived and was brokering the material we need. Aren't Quegans a bit far from home?'

'You could say that,' observed Locklear. 'We ran into some and wondered if they had friends.'

'Well, according to the gossip, they were all heading north

from here, so if you didn't run into a big bunch, they've got friends around.'

Locklear said, 'That's as I figured.'

They ate in silence for a while, as Dubal nursed his mug of ale. Then the dwarf said, 'You wouldn't have run across one of those Armengar monster hunters coming from the north, have you?'

'Monster hunter?' asked Owyn.

Locklear said, 'Beast Hunter, is what he means. I met one, once.' He smiled at the memory. They had been travelling with Prince Arutha away from a band of moredhel, and had run into a Beast Hunter from Armengar with his Beast Hound. It had been a trap, but it had saved them from the pursuing moredhel. 'No, I think those that remain are up in the hills of northern Yabon. Why?'

'Oh, we've got a Brak Nurr loose in the mine and need someone to hunt it down for us. We can either rebuild the mine or hunt the thing, but there aren't enough of us on this side of the mountain to do both.'

'What's a Brak Nurr?' asked Owyn. 'I've never heard of such a creature.'

'It's more a nuisance than a menace,' said Dubal. 'It's a pretty stupid creature, but most of their kind stay in the lower mines and tunnels under the mountain. It's roughly man-shaped, but looks like a walking pile of rocks. That's part of its danger, boy,' Dubal said to Owyn. 'You can't see one until you've stepped on its toes, as often as not. They're slow and lumbering, but they're strong and can crush a man's skull with a single blow. This one came up because of the rockslide, I think, but whatever the cause, it's tried to hurt a couple of our lads. We've chased it off, but can't take the time to hunt it down. If you're up for a bit of fun, I can take you along and if you rid the mines of it, I'll be happy to see you rewarded.'

'Reward?' said Locky. 'That's always a good word, but time doesn't permit. If circumstances bring us to the mines any time soon, we'll be glad to help, but for the moment, we're heading south.'

Dubal stood. 'I understand. Once we get the tunnels finished, we'll go looking for the beastie. Now, I'm for bed and an early start. It was good seeing you again, squire, even in such company as this,' he said, indicating Gorath. 'Good fortune follow you.'

'And you, Dubal.'

Locklear finished eating and rose to approach the innkeeper.

The innkeeper wore a Kingdom-style tunic and trousers, the latter tucked into high-top calfskin boots. But he wore a fur-lined, woven-wool heavy cloak, though it was thrown back, as if even in this warm inn it was too cold for his liking.

'Sir?' asked the innkeeper, his heavy accent making the word sound odd to Locklear.

'Honours to your house,' said Locklear in Tsurani.

The man smiled and said something in return. Locklear smiled and shrugged. 'Sorry, that was all the Tsurani I know.'

The man's smile broadened. 'More than most,' he said. 'You're not from LaMut,' he observed.

'True. I learned a little of your native tongue at Sethanon.'

'Ah,' said the innkeeper, nodding in understanding. Few who were at Sethanon spoke of what had occurred there, mostly because few understood it. At the height of the battle a great upheaval had driven both armies, invaders and defenders, fleeing from the city. A green light from the heavens and the appearance of something in the sky, followed by the destruction of the centre of the city, had rendered most men stunned, and a few deaf, after the battle. No one was certain what had happened, though most conceded a great magic had been unleashed. Most speculated the magician Pug, a friend of the Prince, had a hand in it, but no one seemed to know for certain.

Locklear had missed most of the end of the battle, being hidden in a cellar in the city, but he had heard enough accounts from other eyewitnesses to have formed a pretty clear picture in his own mind. And there was a special bond among those who had survived the Battle of Sethanon, irrespective of their place of birth, for it had been Tsurani, Kingdom, and even Keshian soldiers, who had driven the moredhel and their goblin allies back into the Northlands.

'What I said,' explained the innkeeper, 'was "Honour to your houses, and be welcome to the Blue Wheel Inn".'

'Blue Wheel? That's one of your Tsurani political parties, isn't it?'

The innkeeper's broad face split into a smile, revealing even white teeth. His dark eyes seemed to glint in the lanternlight. 'You do know of us!' He extended his hand, Kingdom fashion, and said, 'I am Sumani. If there is anything that my servants or I may do, you need only ask.'

Locklear shook the innkeeper's hand and said, 'A room for the night after we finish our meal would serve. We have business in the castle tomorrow at dawn.'

The stocky ex-fighter nodded. 'You're in luck, my friend. Last night I would have had to express my regrets and endure the shame of being unable to fulfil your request. We were full, but this morning a large party departed and we have rooms.' He reached under the bar and produced a heavy iron key. 'On my home world this would have been worth a man's life; here it is but a tool.'

Locklear nodded, understanding the scarcity of metals on Kelewan. He took the key. 'Large party?'

'Yes,' said Sumani. 'Foreigners. Quegans, I believe. Their speech was strange to my ear.'

Locklear looked around the obviously prosperous inn. 'How did a Tsurani soldier end up running an inn in LaMut?'

'After the war, Earl Kasumi gave those of us who had been trapped on this side of the rift the opportunity to live as Kingdom citizens. When the rift was reopened, he gave those of us here in LaMut the choice of leaving service and returning to the Shinzawai estates on Kelewan. Most stayed, though some left service and returned to serve again with Kasumi's father, Lord Kamatsu. A few of us, however, retired here in LaMut. I had no living family back home.' He glanced around. 'And to tell the truth, I live better here than I would have back home. There, I might have become a farmer, or a labourer on the Shinzawai estates.' He pointed through the open door to the kitchen to where a tall, stout woman was hard at work preparing food. 'Here, I have a Kingdom wife. We have two children. Life is good. And I am part of the city's militia, so I still train with my sword. The gods of both worlds smile on me and I prosper. I find business to be as challenging as warfare.'

Locklear smiled. 'I have no head for business, though I have been told it often is like warfare. What gossip?'

The old former fighter said, 'Much. Many travellers in LaMut over the last month. Much speculation. A large party of Great Ones came through here last week. And it is rumoured some brigands from my home world, grey warriors, have also been seen near the city.'

'Grey warriors?' asked Locklear. 'Houseless men? What would they be doing here in LaMut?'

Sumani shrugged. 'It may be those without honour have heard that here a man may rise by his own wits and talents, and not be bound by his rank at birth. Or it may be they are seeking riches in this land. With a grey warrior, who can say?' A frown crossed Sumani's face.

'What?' asked Locklear.

'Just this one thing: the rift is controlled by those who serve the Great Ones on Kelewan, and Kingdom soldiers guard the gate on this side. To pass through, these grey warriors would have to have documents, or allies among those guarding the rift gate.'

'Bribes?' asked Locklear.

'Here, perhaps. I've found in the Kingdom the concept

of honour is different than at home. But betrayal from the servants of the Great Ones?' He shook his head. 'That is impossible.'

'Thanks,' said Locklear, smelling a puzzle. 'I'll keep my eyes and ears open.'

The Tsurani laughed. 'That is a funny thing to say,' he observed. 'Let me know if I may be of any further service.'

Locklear nodded. He took a lantern from the innkeeper and returned to the table. Gorath and Owyn rose, and Locklear led his companions up the stairs to a simple room with four beds. He motioned for Owyn to help him move one of the beds across the door, barring it against a sudden attack, then he moved another directly below the window. 'Owyn,' he said, pointing to the bed under the window, 'you sleep there.'

'Why?' asked the young man from Timons. 'It's draughty under there.'

Gorath looked on with a slight turn to his lip, as if amused, as Locklear answered, 'Because if anyone climbs in through the window, they'll step on you and your shouts will alert us.'

Grumbling, Owyn wrapped his cloak tightly around himself and lay down. Locklear indicated one of the beds to Gorath, who lay upon it without comment. Locklear sat on his bed and blew out the flame in the lantern, plunging the room into darkness. Voices from the common room below carried upstairs, and Locklear let his mind wander. The presence of foreigners and the attack by the Quegans worried him, and the rumour of Tsurani grey warriors in the area caused him additional concern, but fatigue and his injury caused him to quickly fall asleep.

TWO

Deception

The soldier waved them in.

'You may enter,' he informed Locklear.

Locklear led his companions into the guardroom of the castle.

They had approached the castle on foot, after an early-morning climb up a long, winding road from the city. He was doubly glad they had chosen to spend the night in the city. His ribs still hurt, but after a night's sleep in a relatively warm bed and two meals he was feeling twice as fit as he had the day before.

The captain of the castle guard looked up as they entered and said, 'Squire Locklear, isn't it?'

'Yes, Captain Belford,' said Locklear, accepting the captain's hand. 'We met when I passed through on my way north a few months back.'

'I remember,' said the captain with a half-hidden grin. Locklear knew the captain must have heard the rumour of the reason for his banishment to the north. 'What can I do for you?'

'I'd like to see the Earl, if he has the time.'

'I'm sure he'd love to see you again, sir, but the Earl's not here,' said the seasoned old fighter. 'He's off on some errand with a troop of men – all Tsurani-bred – leaving me here to take care of things.'

'The Countess?' asked Locklear, inquiring after Kasumi's wife.

'Down in the city, actually. Shopping and visiting with her family.' Earl Kasumi had married the daughter of one of LaMut's more prosperous merchants. 'If you need something official, you can wait until one of them gets back or ask me, squire. As long as you don't need an armed escort somewhere.'

Locklear grimaced. 'I had been thinking about asking for some men to accompany us down to Ylith.'

'Wish I could oblige, squire, and if you've the Prince's warrant with you, I'd scrape together a dozen swords for you, but as it is, the Earl's off training recruits, I've got my usual patrols along the frontier, and the rest of the lads are out looking for a bunch of Tsurani renegades.'

Owyn said, 'Renegades?' Locklear had mentioned nothing of the Tsurani grey warriors to his companions.

'I heard some rumours,' was all Locklear said.

The captain motioned for the three of them to sit. Owyn was left standing when Gorath and Locklear took the only two free chairs in the office. 'I wish it was only rumours,' said Belford. 'You know that Tsurani magician, Makala?'

'By reputation only,' said Locklear. 'He was due to arrive in Krondor a few weeks after I departed some months ago. The other Tsurani Great Ones spoke of him, but as they weren't the most sociable bunch, I only gathered a few things about him. He's very influential in their Assembly of Magicians, is keen to foster trade and what I believe the Prince is calling "cultural exchanges" between the Empire of Tsuranuanni and the Kingdom, and he was personally coming for a visit.'

'Well, he did that,' said the captain. 'He arrived here a few days ago and called on the Earl. Every Tsurani of any rank does that, as the Earl's father is very important on the Tsurani home world. So it's a duty thing.' The old captain rubbed his beard-stubbled chin with a gloved hand. 'The Tsurani are very deep into "duty", I have learned in my time with the Earl. Anyway, they were here for a couple of days, Makala, some other Black Robes, and honour guards and bearers and the bunch, and it seems some of the bearers weren't really bearers, but were some kind of dishonoured warriors from the Empire.'

'Grey warriors,' said Locklear. 'I heard.' That would explain how the grey warriors got through the rift, thought Locklear, disguised as bearers.

'That's who my lads are looking for. Rumour is they fled east. If they get over the mountains and into the Dimwood, we'll never find them.'

'Why the fuss?' asked Owyn. 'Are they slaves or indentured?'

'Squire?' said the captain pointedly.

'He's the son of the Baron of Timons,' explained Locklear.

'Well, young sir,' said the captain, 'these men are something like outlaws on their own world, which by itself isn't enough to have me chasing after them, but here they stole something of value to this Makala – a ruby of some rarity, I gather – and he's making enough of a fuss about it that you'd think the gods themselves lent it to him and he's got to take it back in a week. So the Earl, some because he's polite, and some because he's Tsurani and used to jumping whenever one of those Black Robes barks, he's got us combing the hills looking for those bastards.'

Locklear smiled at Owyn, as if asking if that was explanation enough. The captain looked at Gorath, as if expecting him to say something. Gorath remained silent. Locklear didn't know if the captain recognized the moredhel for what he was or thought him an elf, and didn't see the need to explain things to him. The captain said, 'What would you need an escort for, if I may make so bold as to ask?'

'We've had some problems,' said Locklear. 'Someone's hired Quegan swords to keep us from reaching Krondor.'

The captain stroked his chin again and remained silent a long moment as he thought. 'Here's one thing I can do,' he said. 'I've got to run a patrol out to the border with the Free Cities. I can have you travel with it until it turns westward, almost half-way between LaMut and Zūn. That'll get you part of the way in safety.'

Locklear was silent a moment, then said, 'I have a better idea.'

'What?' asked Captain Belford.

'If you can pick three men to play our parts, and ride conspicuously out the south city gate, we'll head east and slip over the mountains and head south to Krondor along the east mountain highway, where we won't be expected.'

'A ruse?' asked the captain.

'One I learned from the Prince,' said Locklear. 'He used it to good effect in the Riftwar. If you can lead away those looking for us, long enough for us to reach the far side of the mountains, we should be safe.'

'I can arrange that.' He glanced at Owyn and Gorath. 'I've got some men who can pass for you, if we keep the hood up on the one playing your elf friend, here.' He stood up. 'Let me arrange to have the evening patrol stop by your lodgings . . . ?' He looked at them questioningly.

'The Inn of the Blue Wheel.'

Belford smiled. 'Sumani's place. Don't let his smiling countenance fool you; he's a tough boot. If you get the time, have him show you some of his fighting tricks. He'll make time for a few coins. His decision not to stay in service was our loss.'

The captain left and returned a short time later. 'It's taken care of. Head back to the city and let anyone who might be following you see you return. Lie low in the inn until tonight and I'll have three horses waiting for you in the inn's stable.' He handed Locklear a piece of parchment. 'Here's a pass. If one of our lads on the road to the east stops you, this will set him right.'

Locklear rose. 'Thank you, captain. You've been a great help. If there's anything I can do for you when you're next in Krondor, please tell me.'

The old captain smiled. Rubbing his chin once more he

said, 'Well, you could introduce me to that merchant's young wife I hear got you run up this way in the first place.'

Owyn grinned and Gorath remained impassive as Locklear blushed and grimaced. 'I'll see what I can do.' They rose and departed the office.

Owyn said, 'We walk?'

'We walk,' said Locklear as they headed for the main gate of the castle. 'But at least it's downhill.'

Gorath said, 'That is actually more tiring.'

Locklear swore. 'It was a joke.'

Gorath said, 'Really?' His tone was so dry it took a moment for Owyn to realize he was twitting Locklear. Owyn kept his own mirth in check and they started back toward the city.

Locklear slipped through the door into their room. Gorath looked up without alarm, but Owyn jumped off the bed. 'Where have you been?'

'Nosing around. Sitting up here might be smarter, but I've got this itch to scratch.'

Gorath looked on, but still said nothing.

Owyn said, 'Itch?'

Locklear smiled. 'Too many years of keeping the wrong sort of company, I suppose, but the reports of those grey warriors and the theft of some sort of rich item dear to a Tsurani Great One had me thinking. If I stole something on a different world, how would I dispose of it?'

'Depends on what it is, I guess,' offered Owyn.

Gorath gave a slight nod, but still said nothing.

'There would have to be a local contact, someone who knew where one disposes of something of value.'

'And you expect to discover this person in the midst of the throng of this city and use him to trace this band of thieves?' asked Gorath.

'No,' said Locklear waving away the comment. 'The captain said the stolen item is a gem, which being from Kelewan isn't a shock. There isn't much on that world of value that's also easy to transport that would fetch a high value here. So my thinking is that the best way to find this missing gem is to learn where it's most likely to end up.'

'A fence?' asked Owyn.

'No, for if as I suspect the value of the ruby is enough to give a band of desperate men a new start on a strange world, it would have to be the sort of man who has a legitimate enterprise, one likely to mask the movement of this item.'

'You seem to understand this sort of business better than a noble of your race should,' observed Gorath.

'I said I kept the wrong sort of company. After buying a few drinks, I discovered there's a merchant with less than a stellar reputation who deals in gems, jewellery, and other luxury items. He's a man named Kiefer Alescook.'

'Who told you this?' asked Owyn.

'Our host, actually,' said Locklear, motioning it was time for them to depart. They rose and gathered their gear, and moved out down the stairs to the common room. With a wave goodbye to Sumani, they moved through the door. Once outside the inn, Locklear motioned for them to walk around the corner to the stabling yard next to the inn. They moved inside the door and found three men waiting for them, each holding two horses.

One said, 'Switch cloaks, quickly!'

Each was of a like height with Locklear and his companions and the exchange was made. If the man playing the part of Gorath had any notion of whom he was impersonating, he kept such thoughts to himself, merely handing Gorath a large blue cloak, taking the dark grey one worn by the moredhel. The others switched cloaks and Locklear took the reins of one of the horses.

By the time the three impostors were mounted, the sound of hooves on the stones announced the arrival of the patrol that would head down toward Zūn this evening. From outside the gate of the stabling yard, a sergeant shouted, 'We're here to escort you south, Squire Locklear!'

Locklear took his cue and shouted back, 'We're ready!' He nodded to the three men impersonating them who rode off and joined the van of the column. Locklear waited and after a few minutes said, 'Owyn, you ride out, turn left and head straight out the gate. Ride a mile, then wait. Gorath and I will be behind you by a few minutes.'

Gorath grunted his approval. 'So should anyone linger, he won't see three riders.'

Locklear nodded and Owyn said, 'Hold this, please.' He handed his quarterstaff to Locklear, climbed into the saddle, then took the long oaken pole back. With a deft movement, he slung it over his shoulder, through his belt, then twisted it, so it hung across his shoulders and back, not encumbering him or the horse too much.

Gorath easily mounted, though he looked slightly ill at ease.

'Don't ride much?' asked Locklear as Owyn departed.

'Not really. It's been a while, thirty or so years.'

'Not a lot of horses in the Northlands?'

Without bitterness, Gorath said, 'Not a lot of anything in the Northlands.'

Locklear said, 'I remember.'

Gorath nodded. 'We bled at Armengar.'

Locklear said, 'Not enough. It didn't keep you from coming through Highcastle.'

Gorath pointed with his chin. 'We should go now.' He didn't wait for Locklear, but put heels to the sides of his horse and rode out.

Locklear hesitated a moment, then followed after. He overtook the dark elf as he rode easily through the foot traffic of the city. Men hurried home for evening meals while shops closed on every side. Travellers fresh in from the highway hurried toward the inn, eager to wash away the day's trail dust with an ale, and women of the night began to appear on street corners.

Locklear and Gorath rode out the gate, ignored by the

guards, and set their horses to cantering. A few minutes later they spied Owyn sitting on the side of the road.

When they reached him, he turned and said, 'Now what?'

Locklear pointed toward a stand of woods a short distance away. 'A cold camp, unfortunately, but at first light we ride north a few miles. There's a mine road to the east that leads over the mountains. We'll take that, then turn south on the other side. With luck we'll avoid those seeking our friend here and make our way safely to the King's Highway south of Quester's View.'

Owyn said, 'That means we're going to come out near Loriel, right?'

'Yes,' said Locklear, with a smile. 'Which means we'll have the chance to visit one Kiefer Alescook along the way.'

'Why involve ourselves in this matter?' asked Gorath. 'We need to hurry to Krondor.'

'We are, and a few minutes' conversation with Master Alescook may yield us a benefit. Should we discover the whereabouts of this missing gem, we win credit with Prince Arutha, for I am certain he wishes to be a gracious host to the visiting magicians from Kelewan.'

'And if we don't?' asked Owyn as they rode toward the woods.

'Then I still have to come up with a compelling reason why I left Tyr-Sog without his leave and returned with only this moredhel and an unlikely story.'

Owyn sighed aloud. 'Well, you think of one to tell my father when I get back home and I'll try to come up with something to tell the Prince.'

Gorath chuckled at this.

Owyn and Locklear exchanged glances. Locklear shook his head in the evening gloom. He had never considered the dark elves might have a sense of humour.

The wind was cold in the passes, for as winter was coming, in the elevations above them snow already clung tenaciously to the rocks and ice lurked in depressions in the road, making the footing dangerous.

They rode slowly, Locklear and Owyn both with their cloaks pulled tightly around them. Gorath kept his hood up, but rode without apparent discomfort.

'How much longer?' asked Owyn, his teeth chattering.

'A half-hour less than the last time you asked,' said Locklear.

'Squire,' said Owyn. 'I'm freezing.'

Locklear said, 'Really. How unusual.'

Gorath held up his hand. 'Quiet,' he said softly, with just enough authority and volume to carry to his companions, but no farther. He pointed up ahead. 'In the rocks,' he whispered.

'What?' asked Locklear in hushed tones.

Gorath only pointed. He held up four fingers.

'Maybe they're bandits,' whispered Owyn.

'They're speaking my tongue,' said Gorath.

Locklear sighed. 'They're covering all the roads, then.'

'How do we proceed?' asked Owyn.

Pulling his sword, Gorath said, 'We kill them.' He spurred his horse forward, with Locklear hesitating only an instant before following.

Owyn reached up and quickly pulled out his staff, tucking it under his arm like a lance, then urging his horse forward. He heard a shout as he rounded a turn in the trail and entered a widening in the road where one dark elf lay dying in the road as Gorath sped past him.

The other three were not so quickly taken, but rather hurried up into higher rocks where the horses couldn't follow. Locklear didn't hesitate and in a move that startled Owyn, the squire jumped up on his saddle and leaped off the running horse's back, knocking a moredhel from the rock he was climbing.

On his right Owyn saw another one turn, rapidly stringing his bow, then reaching in a hip quiver for an arrow. Owyn urged his horse forward, and swept his staff, striking the bowman below the knee. The bowman went down, his feet shooting out from under him, and struck the rocks with the back of his head.

Owyn's mount shied from the sudden motion near his head and suddenly Owyn found himself falling backwards. 'Ahhhh!' he cried, and then he struck something softer than the rocks. A stunned 'oof' accompanied the impact, and a groan told him he had landed atop the already injured dark elf.

As if scorched by the touch of a flame, Owyn turned over and sat up, scrambling backward. Suddenly he was struck from behind by his horse as the animal turned and sped down the trail. 'Hey!' Owyn shouted, as if he could order the animal to stop.

He then realized there was a struggle going on, and the twice-struck moredhel was attempting to rise. Owyn looked around for a weapon and saw the fallen archer's bow. Owyn grabbed it, and using it like a club, struck the moredhel in the head with as much strength as he could muster. The bow shattered and the warrior's head snapped back. Owyn was certain he wouldn't rise again.

The young magician turned to see Locklear standing away from a now dead dark elf, while Gorath likewise stood over a fallen foe. The moredhel turned and looked in all directions, as if seeking another foe. After a moment, he put up his sword and said, 'They are alone.'

'How can you tell?' asked Locklear.

'These are my people,' said Gorath without apparent bitterness. 'It is unusual for even this many to travel together this far south of our lands.' He motioned toward a small fire. 'They didn't expect to encounter us.'

'Then what were they doing here?' asked Locklear.

'Waiting for someone?'

'Who?' asked Owyn.

Gorath looked around in the late-afternoon light as if

seeing something in the distant peaks, or through the rocks on either side of the trail. 'I don't know. But they were waiting here.'

Locklear said, 'Where is your horse, Owyn?'

Owyn looked over his shoulder and said, 'Back down there somewhere. I fell off.'

Gorath smiled. 'I saw you land on that one over there.' He indicated the body.

Locklear said, 'Hurry back down the trail and see if you can find him. If he's heading back toward LaMut, we'll have to ride in rotation. I don't want to be slowed any more than necessary.'

As Owyn ran off, Gorath said, 'Why don't you leave him behind?'

Locklear studied the moredhel's expression as if trying to read him, then at last he said, 'It's not our way.'

Gorath laughed mockingly. 'My experience with your kind tells me otherwise.'

Locklear said, 'Then it's not *my* way.'

Gorath shrugged. 'I can accept that.' He set to examining the corpse at Locklear's feet and after a moment said, 'This is interesting.' He held out an object for Locklear's examination.

'What is this?' asked Locklear, looking at a multi-faceted stone of an odd blue hue.

'A snow sapphire.'

'Sapphire!' said Locklear. 'It's as big as an egg!'

'It's not a particularly valuable stone,' said Gorath. 'They are common north of the Teeth of the World.'

'So it's, what? A keepsake?'

'Perhaps, but when a war party leaves our homeland, we travel light. Weapons, rations, extra bowstrings, and little else. We easily live by forage.'

'Maybe this isn't a war party,' suggested Locklear. 'Maybe they live around here?'

Gorath shook his head. 'The last of my people south of the

Teeth of the World lived in the Grey Towers and they fled
to the Northlands with the coming of the Tsurani. None of
my race has lived this near the Bitter Sea since before the
Kingdom came to these mountains. No, while not of my
clan, these are from the Northlands.' He put the gem in his
belt pouch and continued to examine the bodies.

Time passed and finally Owyn put in an appearance,
leading his horse. 'Damn all horses,' he swore. 'He made
me chase him until he got bored.'

Locklear smiled. 'Next time, don't fall off.'

'I didn't plan on it this time,' said Owyn.

Gorath said, 'We need to hide these.' He pointed to the
four dead moredhel. He picked up one and carried it a short
way down the trail then unceremoniously threw the corpse
over the side of a ravine.

Owyn looked at Locklear, and the young magician tied his
horse's reins to a nearby bush. He picked up the feet of the
nearest corpse while Locklear lifted the creature under the
shoulders.

Soon all four bodies were consigned to the ravine hundreds of feet below. Locklear mounted as did Gorath and
Owyn. Leaving for the time being the mystery of why these
moredhel were waiting at this lonely spot on a rarely used
trail, they rode on.

Loriel appeared before them, a small city – really a large town
– nestled into the large valley which ran eastward. Another
valley intersected from the south.

Gorath said, 'We need food.'

'A fact of which my stomach is well aware,' answered
Locklear.

Owyn said, 'Not that I'm in a hurry to face my father, but
this is turning into a roundabout journey, squire.'

Locklear pointed to the southern valley. 'There's a road
through there that's a very straight course to Hawk's Hollow.
From there we have our choice of routes, south along a

narrow ridge trail, or southwest back to the King's Highway.'

Gorath said, 'And then to Krondor?'

'And then to Krondor,' agreed Locklear. 'Something in all this is making what my friend Jimmy calls his "bump of trouble" itch like I've been attacked there by fleas.

'Gorath, this stolen ruby, the Tsurani magicians, all of it is somehow . . . more than coincidence.'

'How?' asked Owyn.

'If I knew,' said Locklear, 'we wouldn't be stopping off to visit Mr Alescook. He may know something or know someone who knows what it's about, but the more I think on this mystery, the more it bothers me that I don't know what's behind all this.

'But we're going to find out or die trying.'

Owyn didn't look happy at the second choice, but said nothing. Gorath just looked out over the town as they rode down towards a small guard post that sat beside the trail.

A town constable of advancing years and considerable girth held up his hand and said, 'Halt!'

The three reined in and Locklear inquired, 'What is it?'

'We've had a rash of renegades around here, lately, m'lad, so state your business.'

'We're travelling south and stopping for provisions,' said Locklear.

'And who might you be, to be riding down out of the mountains?'

Locklear produced the paper given him by Captain Belford and said, 'This should explain as much as you need to know, constable.'

The man took the document and squinted at it. Locklear realized he couldn't read, but he made a show of studying it. Finally, convinced by the large embossment at the bottom, the constable handed back the paper and said, 'You may pass, sir. Just be wary if you're out after dark.'

'Why?' asked Locklear.

'As I said, sir, lots of ruffians and bandits passing by lately, and not too few of those murderous Brothers of the Dark Path. Look a bit like your elf friend there, but with long black nails and red eyes which shine in the night.'

Locklear could barely hold back his amusement as he said, 'We'll be wary, constable.'

They rode past and Gorath said, 'That one has never seen one of my people in his life.'

'So I gathered,' observed Locklear, 'though I must pay more attention to your eyes at night. I may have missed the red glow.'

Owyn chuckled and they found themselves an inn. It was dirty, crowded and dark, which suited Locklear fine as he was low on funds. He had thought about asking Captain Belford for a loan, but decided the captain's only response would have been, 'wait for Earl Kasumi,' and while Locklear didn't mind taking a circuitous route to get to Krondor to avoid ambushes, he was anxious to put the mystery of what was occurring in the Northlands before Arutha.

There were no rooms available, a situation that surprised Locklear, but the innkeeper gave them leave to sleep in the commons. Owyn grumbled at the need, but Gorath kept his thoughts to himself.

So far no one had objected to the moredhel's presence along the way, either because they didn't recognize him for what he was, mistaking him for an elf, or because a moredhel with renegade humans in these mountains was not all that unusual a sight. Whatever the cause, Locklear was grateful he didn't need to deal with curious onlookers.

They ate at a crowded table, and after the meal listened to an indifferent troubadour. There were some games of chance and Locklear itched to try his hand at some cards, either pashawa or pokir. He resisted the impulse, as he could ill afford to lose, and one lesson taught him by his father and older brothers was don't gamble what you can't afford to lose.

As the inn settled down and those sleeping in the commons began to claim corners and places under tables, Locklear approached the barkeep, a heavy-set man with a black beard. 'Sir?' he asked as Locklear moved between two other men to stand before him.

'Tell me, friend,' began Locklear. 'Is there a merchant in this town who deals in gems?'

The barkeep nodded. 'Three doors down on the right. Name's Alescook.'

'Good,' said Locklear. 'I need to purchase a gift for a lady.'

The barkeep grinned. 'I understand, sir. Now, one word: caution.'

'I don't understand,' said Locklear.

'I'm not saying Kiefer Alescook can't be trusted, but let's just say the source of some of his merchandise is a bit dodgy.'

'Ah,' said Locklear, nodding as if now he understood. 'Thanks. I'll bear that in mind.'

Locklear returned to the table and said, 'I've found our man. He's nearby and we'll see him first thing in the morning.'

'Good,' said Gorath. 'I tire of your company.'

Locklear laughed. 'You're not exactly an ale and fair song yourself, Gorath.'

Owyn said, 'Well, whatever. I'm tired and if we're to sleep on the floor, I don't want to get too far from the fire.'

Locklear realized that men were now bedding down for the night and replied, 'Over there.'

They moved to the indicated spot and unrolled their bedding. After a few minutes of listening to the sounds of hushed conversation from those few men still at the tables or the door opening and closing as men left to return to their homes, Locklear fell into a deep sleep.

The merchant looked up as the three men entered the room.

He was an old man, looking frail to the point of infirmity. He regarded the three with rheumy eyes. He studied Gorath for a moment, then said, 'If you've come for gold, I sent it north with one of your kind two days ago.'

Gorath said, 'I did not come for gold.'

Locklear said, 'We came looking for information.'

The merchant fell silent. After a moment, he said, 'Information? Find a rumour-monger. I deal in gems and other fine items.'

'And from what we hear, you're not too particular as to the source of those items.'

'Are you suggesting I deal in stolen property?' demanded the old man, his voice rising.

Locklear held up his hand. 'I suggest nothing, but I am seeking a particular stone.'

'What?'

'A ruby, unusual in size and character. I seek to return it to its rightful owner, no questions asked. If you came by it, no fault will be placed at your feet, *if* you help us recover it. If you don't, then I suggest you may receive a visit from a royal magistrate and some very disapproving guardsmen from the garrison at Tyr-Sog.'

The old man's expression turned calculating. His balding pate shone in the light of a single lantern that hung overhead. With feigned indifference he said, 'I have nothing to hide. But I may be able to help you.'

'What do you know?' asked Locklear.

'Lately, my business has been brisk, but it's an unusual sort of trade, and I've been in this business for fifty years, lad.

'Recently, I've been handling transactions for parties I have not met, through agents and couriers. Most unusual, but profitable. Gems of high quality, many of them very rare, even remarkable, have passed through my hands.'

'Tsurani gems?' asked Locklear.

'Precisely!' said the old man. 'Yes, similar enough to our own rubies, sapphires, emeralds and the like to be recognized as such, but with slight variations only an expert might notice. And also, other gems unlike any found on this world.'

'Whom do you represent?' asked Locklear.

'No one known to me,' said the old man. 'At irregular intervals of late, dark elves like your companion have come here, and they drop off gems. Later a man comes from the south and brings me gold. I take a commission and wait for the dark elves to return and take the gold.'

Gorath turned to Locklear. 'Delekhan. He's using the gold to arm our people.'

Locklear held his hand up, requesting silence. 'We'll talk later.' To the old man he said, 'Who buys the gems?'

'I don't know, but the man who receives them is known as Isaac. He lives down in Hawk's Hollow.'

'Have you seen this Isaac?' asked Locklear.

'Many times. He's a young man, about your height. Light brown hair he wears long to his shoulders.'

'Does he speak like an Easterner?'

'Yes, now that you mention it. He sounds court bred at times.'

Locklear said, 'Thank you. I will mention your aid should any official investigation come of this.'

'I am always eager to help the authorities. I run a lawful enterprise.'

'Good.' Locklear motioned toward Gorath's purse and said, 'Sell him the stone.'

Gorath took out the snow sapphire he had taken from the dead moredhel and put it down before Alescook.

The merchant picked it up and examined it. 'Ah, a nice one. I have a buyer for these down south. I'll give you a golden sovereign for it.'

'Five,' said Locklear.

'These are not that rare,' said Alescook, tossing it back to Gorath, who started to put it away. 'But, on the other hand . . . two sovereigns.'

'Four,' said Locklear.

'Three, and that's done with it.'

They took the gold, enough for a meal along the way, left and went outside. To his companions Locklear said, 'We're passing through Hawk's Hollow on our way to Krondor, so our next choice is easy. We find Isaac.'

As he mounted his horse, Gorath said, 'This Isaac is known to you, then?'

Locklear said, 'Yes. He's the second biggest rogue I've known in my life. A fine companion for drinking and brawling. If he's caught up in something dodgy, it wouldn't surprise me.'

They turned their horses southward and left the large, rolling valley of Loriel, entering the narrow river valley leading southward. Locklear had been able to purchase a little food at the inn, but the lack of funds was starting to worry him. He knew they could hunt, but his sense of something dark approaching was growing by the day. A renegade moredhel chieftain bringing warning of possible invasion, money moving to the north to buy weapons from weapons runners, and somehow the Tsurani were involved. Any way he looked at this, it was a bad situation.

Unable to put aside his foreboding, he kept his thoughts to himself.

Gorath held up his hand and pointed. Softly he said, 'Something there.'

'I don't see anything,' said Owyn.

'If you did, I would not need to warn you,' suggested the dark elf.

'What do you see?' asked Locklear.

'An ambush. See those trees. Some lower branches have been hacked off, but not by a woodsman's axe or saw.'

'Owyn,' Locklear asked, 'can you still do that blinding trick?'

'Yes,' said Owyn, 'if I can see the man I'm trying to blind.'

'Well, as we're sitting here, pointing at them, I expect in a moment whoever's behind that brush is going to figure out we've spotted their ambush – '

Locklear was interrupted by six figures rushing forward from the brush on foot. 'Moredhel!' shouted Locklear as he charged.

He felt the sizzling energy speed past him as Owyn sought to blind an advancing dark elf. The spell took effect, for the creature faltered, reaching up to his eyes in alarm.

Locklear leaned over the neck of his horse as an arrow flew past him. 'Get the bowman,' he shouted to Owyn.

Gorath shouted a war cry and rode down one attacker while slashing at a second. Locklear engaged a dark elf who seemed indifferent to facing a mounted opponent, and Locklear knew from bitter experience how deadly the moredhel could be. While rarely mounted themselves, they had faced human cavalry for hundreds of years and were adept at pulling riders from horseback. Knowing their tactics, Locklear spurred his mount suddenly, turning it hard to the left. This knocked back the attacker he faced and revealed the one poised to leap and drag him down. Locklear slashed out with his sword, taking the creature in the throat, above his metal breastplate. Locklear kept his horse circling, so he quickly faced his first attacker.

The sizzling sensation told him Owyn was once more blinding an opponent, and Locklear hoped it was the bowman. The moredhel who had fallen back as the horse spun pressed forward with a vicious slash at Locklear's leg.

He barely got his sword down in time and felt the shock run up through his arm. His stiff ribs hindered his parry and the flat of his own blade slammed into his horse's side, causing the animal to shy.

Locklear used his left leg and moved the animal back into a straight line, twisting his body to keep his eyes upon his foe. His ribs hurt from the effort, but he stayed alive as the moredhel swung at him again. He knocked that blow aside and delivered a weak counter which slapped his opponent in the face, irritating him more than doing any real damage.

But the blow did slow the moredhel's advance, and Locklear got his horse turned to face his foe. Locklear remembered something his father had drilled into him and his brothers; a soldier who has a weapon and doesn't use it is either an idiot or dead.

His horse was a weapon, and Locklear put his legs hard against his horse's flanks and tugged hard on the reins with his off hand. The horse picked up a canter, and to the moredhel it was as if the horse suddenly leaped at him.

The warrior was a veteran and dodged to one side, but Locklear reined his horse in, turning hard to the left. To the moredhel, it looked as if Locklear was turning away, and the creature pressed forward.

Locklear kept the horse turning in a tight circle, and suddenly the moredhel realized his error as the young squire completed his circle with a slashing downward blow. This was no irritating tap, but a powerful blow which smashed bone as it cut into the side of the moredhel's skull.

Locklear glanced toward Gorath and saw him beset by two foes, then looked back to Owyn, and saw that he was on foot a hundred yards away and holding a swordsman at bay with his staff. Hoping the bowman was still blinded by Owyn's magic, Locklear rode to Owyn's rescue.

He kicked hard at his horse's flanks and the animal leaped forward so that he was approaching at a gallop when the moredhel heard him coming. The dark elf turned to look at his second opponent, giving Owyn the opening to strike with the butt of his staff. He broke the creature's jaw and sent him slumping to the ground.

Locklear reined his horse in so suddenly the animal planted his hooves and almost sat. Spinning the horse around, Locklear waved to Owyn, shouting, 'Keep the bowman off us!'

As if the Goddess of Luck had turned a deaf ear to him,

Locklear was lifted out of the saddle by an arrow. He struck the ground hard, barely avoiding broken bones by rolling. The arrow in his left shoulder snapped and the pain caused his vision to swim and took his breath away.

For the briefest instant, Locklear fought to keep conscious, then he felt his eyes focus and he willed away the pain in his shoulder. A strangled cry behind him made him turn. Over him reared a moredhel, sword raised to strike. Suddenly Gorath was behind the moredhel, and he plunged his sword into the moredhel's back.

Owyn ran past, wheeling his staff above his head. Locklear looked up as his would-be killer fell to his knees, then keeled over. Gorath turned before Locklear could speak and ran after Owyn.

Locklear rose slowly on wobbly legs as he saw Owyn rush forward and strike a moredhel bowman who was vainly rubbing his eyes as if trying to clear them. The bowman was clubbed to his knees, and died a moment later as Gorath delivered the killing blow.

Gorath spun around in a circle once, as if seeking another enemy, but Locklear saw the six were dead. Gorath stood with his sword in hand, frustration on his face, then he shouted in rage. 'Delekhan!'

Locklear stumbled to the dark elf and said, 'What?'

'They knew we were coming!' said Gorath.

Owyn said, 'Somehow they got word south?'

Gorath put up his sword. 'Nago.'

'What?' asked Locklear.

'Not what, *who*,' said Gorath. 'Nago. He's one of Delekhan's sorcerers. He and his brother Narab served the murderer. They are powerful chieftains in their own right, but right now they're doing Delekhan's bidding. Without their help, Delekhan never would have risen to power and overthrown the chieftains of the other clans. Without their help, these – ' his hand swept in a circle, indicating the dead moredhel ' – would not be here waiting.' He knelt next to one of the dead and said, 'This was my cousin, my kinsman.' He pointed to another one. 'That one is from a clan that has been sworn enemy to mine for generations. That they are both serving this monster hints at his power.'

Locklear indicated his shoulder and sank to the ground. Owyn examined it and explained, 'I can get the arrowhead out, but it's going to hurt.'

Locklear said, 'It already hurts. Get on with it.'

While Owyn ministered to Locklear, Gorath said, 'Nago and Narab both have the power of mind speech. Especially with one another. Those we killed on the road to your town of Loriel, or another who spied us, must have passed word to one of the brothers. He in turn alerted these as to our whereabouts.'

Locklear said, 'So the chances are good that before they died, one of these also let Nago know we are here?'

'Almost certainly.'

'Wonderful,' said Locklear through gritted teeth as Owyn used his dagger to cut out the arrowhead. His eyes teared and his vision swam again for a moment, but by breathing slowly and deeply he kept conscious.

Owyn dusted the wound with a pack of herbs from his belt pouch then placed a cloth over it. 'Hold this here; press hard,' he instructed. He went to the nearest body and robbed it of a strip of cloth, cut away with his dagger, then returned to bind it tightly around Locklear's shoulder. 'Between that wound to your ribs and this shoulder, your left arm is close to useless, squire.'

'Just what I wanted to hear,' said Locklear as he tried to move his left arm and found Owyn's observation correct. He could move it scant inches before pain made him stop the attempt. 'Horses?'

'They've run off,' said Owyn.

'Wonderful,' said Locklear. 'I was knocked out of the saddle, what's your excuse?' he demanded of the other two.

Gorath said, 'Fighting on the back of the beast was too awkward.'

Owyn said, 'I can't cast a spell from the saddle. Sorry.'

Locklear stood. 'So we walk.'

'How far is it to Hawk's Hollow?' asked Owyn.

'Too far,' said Locklear. 'If they're waiting for us, much too far.'

THREE

Revelation

The sentry blinked in surprise.

One moment the approach to the town was empty, the next three figures were standing before him. 'What?' he exclaimed, bringing his old spear to something resembling a stance of readiness.

'Easy, friend,' said Locklear. He leaned upon Owyn's shoulder and looked as if he was close to death. They had encountered three more ambushes between the one where their horses had fled and Hawk's Hollow. They had managed to avoid the first two, sneaking around human bandits. The last had been a squad of six moredhel who had been too alert. The fight had been bloody and costly. Gorath was wounded, a nasty cut to his left shoulder that Owyn had barely been able to staunch. Locklear had been injured again, nearly dying if not for Owyn's intervention, and the young magician himself was sporting a half-dozen minor wounds.

'Who are you?' asked the confused sentry. He was obviously a farmer or worker from town, part of the city's militia Locklear guessed.

'Locklear, squire of the Prince's court in Krondor, and these two are my companions.'

'You look like brigands, to me,' replied the guardsman.

'We have proof,' said Locklear, 'but first I'd like to find someone who can help us before we bleed to death.'

'Brother Malcolm of the Temple of Silban is in town, down at Logan's Tavern. He comes through here every six months or so. He'll help you out.'

'Where is Logan's?' asked Owyn as Locklear seemed about to lapse into unconsciousness.

'Just down the street. Can't miss it. Sign out front of a dwarf.'

They made their way to the indicated establishment, which showed a faded sign of a comically drawn dwarf, obviously once painted with vivid colours.

They went inside and found several townspeople sitting by, waiting for a priest in the robes of the Order of Silban who was in the corner ministering to a sick child. A couple of local workers were waiting, one with a bandaged hand, the other looking pale and weak.

The priest looked up as he finished with the boy, who leaped down from his mother's lap without prompting and raced for the door. The priest looked at Locklear and said, 'Are you dying?'

'Not quite,' answered the squire.

'Good, because these fellows were here first and I'll only make them wait if you're near death.'

Mustering as much dry wit as he could under the circumstances, Locklear replied, 'I'll try to let you know when I'm about to die.'

Gorath's patience vanished. He moved to confront the priest and said, 'You will see my companion now. These others can wait.'

The glowering dark elf towered over the small priest and his expression and voice left no room for argument this side of violence. The priest looked once more at Locklear and said, 'Very well, if you think it urgent. Bring him over to this table.'

They half-carried Locklear to the table and laid him out on it. The priest said, 'Who bandaged this?'

'I did,' said Owyn.

'You did well enough,' said the priest. 'He's alive, so that counts for much.'

After Locklear's tunic and the bandages were removed,

the priest said, 'Silban preserve us! You've got three wounds fit to fell a bigger man.' He sprinkled a powder on the wounds, which brought a gasp of pain from Locklear, then the priest began a chant and closed his eyes.

Owyn felt power manifest in the room and the hairs on the back of his neck stood up. He had only been exposed to a little clerical magic in his life and it always seemed odd and exotic to him.

A faint glow from the priest's hands threw illumination over Locklear's wounds and as Brother Malcolm droned his chant, Owyn could see the wounds begin to heal. They were still visible, but no longer fresh and angry. When the priest stopped, they looked old, past the danger stage. The priest was pale from the exertion when he stopped. He said, 'That's all I can do now. Sleep and food will do the rest.' Looking at Owyn and Gorath, he asked, 'Do you have wounds, as well?'

'We do,' said Gorath. 'But we can wait until you tend to those two.' He pointed to the two locals waiting for treatment.

Malcolm nodded. 'Good.' As he moved past Gorath, he said, 'Your manners may be in question, moredhel, but your instincts serve you well. He might have bled to death had we waited another hour.'

Gorath remained silent in the face of being recognized for what he was. He moved to sit next to Owyn and wait.

When the two farmers, one with a smashed finger courtesy of a badly-aimed hammer and the other with a bad case of fever, were finished, Malcolm turned to Gorath and Owyn. 'Who's next?'

Gorath indicated Owyn and the magician went to sit before the priest. He watched with interest as the priest quickly treated and bound his wounds. They spoke little, for Owyn was almost out on his feet.

When Gorath replaced him before the priest, the dark elf said, 'You recognize my race, yet you do not call for the town guard. Why?'

The priest shrugged as he examined Gorath's wounds. 'You travel with men who do not look like renegades to me. You are not here killing and burning, so I assume your mission a peaceful one.'

'Why do you assume I have a mission?' asked Gorath.

'Why else would you travel in the human world?' Malcolm asked rhetorically. 'I have never known the moredhel to travel for pleasure.'

Gorath grunted, forgoing comment.

Malcolm was quickly done and said, 'You should have come second; this wound was more severe than your friend's. But you'll live.' He washed his hands and dried them with a towel. 'It is my mission to aid and serve, but it is custom that those served donate.'

Gorath indicated Locklear, who was now sitting upright at the table upon which he had lain. Locklear said, 'Brother, I fear I may only give you a scant token of our debt, but should you come to Krondor any time soon, visit me and I will repay you tenfold.'

Locklear dug into his purse and judged how much he would need for a room that night, and other costs, then drew out a golden sovereign and two silver royals. 'It is all we can spare.'

'It will do,' said the priest. 'In Krondor, where might I find you?'

'At the palace. I am one of the Prince's men. I am Squire Locklear.'

'Then I shall call upon you when next I'm in Krondor, young squire, and you can settle accounts with me then.' Glancing at Locklear's freshly-bound wounds, he said, 'Go easy on those cuts for another day. By tomorrow you'll feel better. If you avoid being stabbed again any time soon, you'll feel like your old self by week's end. Now, I must go rest. This is more healing in one afternoon than I usually experience in a week.'

The priest left and Locklear slowly rose to cross to the bar and found the innkeeper cleaning up. The portly man said, 'Welcome to The Dusty Dwarf, my friends. What may I do for you?'

'Food and a room,' said Locklear.

They returned to a table and the innkeeper followed soon after, putting down a large platter of cold meats, breads baked earlier that morning, cheese and fruits. 'I've got some hot food cooking for later this evening, but this early in the day, cold fare is all I have.'

Owyn and Gorath were already stuffing food into their mouths as Locklear was saying, 'That will be fine. Some ale, please.'

'Right away.'

The man was back with the ale in a moment, and Owyn asked, 'Sir, what is the story behind the name of this place?'

'The Dusty Dwarf?' said the man.

'Yes.'

'Well, truth to tell, it's not much of a story. Man named Struble owned this place. Called it The Merry Dwarf. Don't know why. But it had a bright sign. He never had the sign repainted in all the years he owned the place, so by the time I bought it from him, the sign was badly faded. All the locals called it The Dusty Dwarf by then, so I just went along. Saves me the cost of getting the sign painted, too.'

Owyn smiled at the story, as the barkeep hurried off to meet the demands of another customer. Locklear looked nearly asleep as he said, 'All right. We have two choices. We can take the main road down to Questor's View, or the back way through Eggly and Tannerus and lose a few days.'

Owyn said, 'I'm only guessing, but from what Gorath has said, this Nago or Narab is keeping in contact with their agents by mind speech. As I said before, I know only a little about this speech, but what I do know is it can be very taxing. The magician Pug's daughter is known to be among the most gifted in the world at this and can speak across vast distances, but she is rare, even unique. For lesser magicians, it requires much rest.'

Gorath looked on impassively, but Locklear said, 'Come to the point, if you don't mind. I'm having trouble staying awake.'

'The point is whoever this magician is, he's lying low in one place, probably guarded, and probably has one or two key agents in a given area. The rest of his orders are being run by messengers, I'm thinking. So they know where we've been, and may have even guessed where we are today, but they don't know which way we'll be going.'

Locklear said, 'Fine, but what does that mean about our choice of route?'

Gorath said, 'It means he must spread his men equally between the two routes, so the best solution is to take the route where we will be best able to defend ourselves or travel with a larger band, such as a trading caravan.'

Locklear motioned to the innkeeper, who came and gave him a key, indicating the room at the top of the stairs. As they mounted the stairs, Locklear observed, 'If we were trying to come back from Kesh, a caravan might be a good cover, but as the King's Highway is usually well patrolled, most traders feel comfortable travelling with a few mercenary guards or none at all. Most commerce along the coast is by ship.'

As they reached the room, Owyn said, 'Could we make for Questor's View and hire a ship?'

'With what?' asked Locklear. 'Captain Belford's letter of introduction isn't exactly the King's writ. If a fleet ship is at anchor, I know I could talk our way aboard and get it bound for Krondor, but I'm not anxious to sit around waiting for one to show up. I'm not anxious for anything but a good night's sleep, finding Isaac and getting this riddle of a special ruby solved, and then figuring out how to get to Krondor as fast as we can.'

Owyn said, 'I can't argue about that night's sleep.'

Gorath said nothing.

*   *   *

An hour after dawn they left the inn and Locklear felt remark-
ably recovered. Where searing agony had accompanied his
every movement the day before, he now only felt slightly
stiff and weak.

He indicated a journey toward the north end of the town
as he said, 'If I know Isaac, he's probably staying at the house
of his cousin, a certain young gentleman named Austin
Delacroix.'

'From Bas-Tyra?' asked Owyn as they started up the busy
street. Windows were opening as vendors put out their
wares for display, or housewives opened up their homes
to the morning air and sun.

'Originally,' said Locklear. 'A family of marginal nobility,
descended from a one time hero of some forgotten war
when Bas-Tyra was a city-state; their house rank is all based
upon that.'

'Your human issues of rank and status are . . . difficult to
understand,' observed Gorath.

'Why?' asked Owyn. 'Don't you have chieftains?'

'We do,' said Gorath. 'But it is a rank earned by deeds,
not one conferred by birth. Delekhan rose by betrayal
and bloodshed, yet he was sheltered by his early service
to Murmandamus and Murad.' He almost spat the last
two names. 'If his son Moraeulf gains his ambition to
inherit from his father, it will be over the bodies of
many such as I. In better times, he would be a valued
sword against our people's enemy, but these are not
better times.'

'This is the house, I think,' said Locklear, pointing to a
once-prosperous dwelling fallen on hard times. The house,
like those on either side, was a small but well-built structure
of wood and stone, with a sturdy door and shuttered windows. But while the others were clean and recently painted,
this was faded and dirty.

Locklear knocked loudly and after a few minutes a sleepy
voice from the other side of the door said, 'What?'

'Isaac?' shouted Locklear, and the door opened.

A young man with light brown hair stuck his head out the door and said, 'Locky?' The door opened wide and the young man bid them enter. He wore only a rumpled tunic and trousers, obviously having slept in them. 'I was just getting up,' he said.

'Right,' said Locklear, as if humouring him.

The room was dark, with the shutters and sashes still closed, and the air was stale. Old food odours and sweat mixed with the sour aroma of spilled ale. The furniture was simple, one wooden table with four chairs, a single shelf behind the table, and another small table upon which a lamp rested. Stairs led to a sleeping loft above. A faded tapestry, once residing in surroundings far finer than those in which they hung now, was the sole item of any note. It hung behind Isaac, framing him with a tableau of a meeting between princes who were exchanging gifts while notables of that day looked on from all sides.

'Locklear,' said Isaac, as if savouring the name. 'What a pleasure. You're wearing your years well. I like the moustache. You always could manage the flamboyant.' He turned away and moved with a visible limp. 'Sit down. I would offer you tea or coffee, but my cousin is temporarily visiting other relatives in Bas-Tyra, and I have just arrived last night, so we are not well provisioned.'

'That's all right,' said Locklear. 'How long's it been? Since Arutha's wedding?'

Isaac sat in a small wooden chair, and crossed his legs so that he kept his weight on his good leg. 'The very day. You should have heard the fit old Master of Ceremonies deLacy threw when he found out I wasn't the Baron of Dorgin's son.'

'That's because there is no Baron of Dorgin,' supplied Locklear. 'If you'd done your research, you would have avoided that gaffe.'

'How was I supposed to know the lands outside the

dwarven enclave are the province of the Duke of the Southern Marches?'

'Study?' suggested Locklear.

'Never my strong suit,' said Isaac with a wave of his hand.

'Well, at least deLacy was too busy with the wedding to toss you out until the next day,' said Locklear. 'We had a good time that night. What have you been doing since?'

'I spent some time in the east with my family, then returned a few years ago to the west. Since then I've been doing odd jobs along the border. So, what brings a member of Krondor's court so far from home with such unusual company?'

'Certain doings, some bloody, which unfortunately point to you.'

'Me?' said Isaac. 'You're not serious.'

'I'm as serious as a royal torturer, Isaac, and you'll have a chance to make a first-hand comparison if you don't answer me truthfully. I'll have Gorath sit on you while I go fetch the local constable. We can have a pleasant talk here, or a very unpleasant one in Krondor.'

Locklear had no intention of summoning the local constable and trying to sort out his claim of rank and authority, especially with no royal writs or warrants. But Isaac didn't know that, and Locklear wasn't about to enlighten him.

'I have no idea what you're talking about,' said Isaac, starting to slowly rise.

Gorath said softly, 'Reach for that sword behind you and you'll have a leg to match the other before your fingers touch the hilt, human.'

'Damn,' said Isaac quietly, sitting back down in the chair.

'The ruby,' said Locklear.

'What ruby?' said Isaac.

'The one you bought from Kiefer Alescook. The one you paid for with gold heading north to buy Delekhan weapons.

The ruby stolen from an important Tsurani magician. The ruby that's the latest in a series of such transactions.'

Isaac ran a hand over his face and back through his hair. 'Locky, it's been hard.'

Locklear's expression turned dark and his voice took on a menacing tone that had Owyn sitting back in surprise. 'As hard as treason, Isaac? As hard as the jerk at the end of a hangman's rope?'

'Who said anything about treason, Locky?' Isaac's manner turned to pleading. 'Look, we were boyhood friends before I had my accident. If our positions had been reversed, you'd know; you'd understand what it's like to be a hired sword with a bad leg. Locky, I was nearly starving when this opportunity came along. I was too far in before I discovered who was behind it.'

'Tell us what you know and I'll do you a favour,' said Locklear.

Isaac looked downfallen, and said in a contrite fashion, 'I was in over my head before I knew who I was dealing with. Alescook is an old acquaintance. I know that from time to time he "finds" gems and jewellery that has . . . ah, "clouded" title is a polite way of putting it.'

'Stolen,' said Locklear.

Isaac squirmed. 'Whatever the cause, the market in the Kingdom is difficult, so those gems find their way south, to Kesh or over the water to Queg or the Free Cities. I'm just a middleman, someone who can take a little trip down to the Vale or over to Krondor or Sarth and put something on a ship. That's all.'

'The ruby?' said Locklear.

Isaac started to rise and hesitated as Gorath leaned forward, hand on the hilt of his sword. Isaac continued rising slowly, then mounted the stairs to the loft above. Locklear motioned with his head to Owyn, who stood up and hurried through a small door in the wall next to the tapestry. He found himself in a tiny kitchen, one dirty enough he would have to be far hungrier than he presently was to consider eating anything prepared there. He ducked through the back door and looked up at a window above, where he saw the head of Isaac disappear back inside. Owyn smiled; Locklear's instincts had been correct. The lame ex-fighter might attempt to escape from a first storey window, but he knew he wasn't quick enough to pull off his escape if someone was waiting below.

A moment later, Locklear called for Owyn's return and the young magician complied. He entered the room and stopped. The hairs on his arm stood up and he said, 'Let me see the stone.'

Isaac handed it to him and said, 'It's really not a very valuable item, but I get paid well.'

Owyn replied, 'I don't know anything about stones and their worth, but I know this one is more than it appears to be.' He looked at it closely. 'This ruby has been prepared.'

'Prepared for what?' asked Locklear. 'Jewellery?'

'No, as a matrix of some kind for magic. I don't know much about this sort of thing.' He put the stone down. 'Truth to tell, I don't know much about any sort of thing magical, which is why I left Stardock. The only magic I've learned so far was from a field magician named Patrus, a sour old character. But my father objected and last I heard Patrus headed north – ' He shook himself out of his reverie. 'It doesn't matter, but what he told me is that some magic is harmonic and can be focused by gems. Or stored in them. He claimed once that magic itself might exist in gem form under the right conditions. For example, you can rig a trap with certain gems, so that whoever steps into a given area is imprisoned.'

'Can you tell what this was used for?'

'No,' said Owyn with a shake to his head. 'It may be something that will be used in the future.'

'So you think it important?' asked Gorath.

'I can now see why the Tsurani magician was so angry about its disappearance.'

Locklear picked up the stone and tossed it in the air a couple of times while he was thinking. After a moment he put away the stone and turned to Isaac. 'Tell us what else you know.'

Isaac looked defeated and said, 'Very well. The stones come through the rift on an irregular basis. Sometimes a bunch, sometimes a single one like this one. Money comes to me in Krondor by various means; never the same twice. There's a new gang in Krondor, run by someone calling himself the Crawler, and he's causing the Mockers fits.'

'Mockers?' asked Gorath.

'Thieves,' said Locklear. 'I'll explain it later. Go on,' he said, looking at Isaac.

'Someone in Krondor is paying for gems. The Tsurani bring them in and hand them over to the moredhel. They run them over to Alescook and I go get them and bring them to Krondor. It's a fairly simple arrangement.'

'But someone's running this. Who and where?'

Isaac sighed. 'There's a village south of Sarth. Called Yellow Mule. Know it?'

'Villages like that don't put up signs, but if it's on the King's Highway, I've ridden through it.'

'It's not. About twenty miles south of Sarth there's a fork in the road, and if you go inland, you're heading toward an old trail up into the mountains. About five miles along that road is where you'll find Yellow Mule. It's why the moredhel are using it. No one travels through there, and it's easy for his kin – ' he indicated Gorath with a jerk of his chin ' – to get there without being seen.

'There's an old smuggler turned farmer named Cedric Rowe now living there. He knows nothing of loyalty to anyone, or anything but gold. He rents out his barn to a Dark Brother named Nago.'

'Nago!' said Gorath. 'If we take him, then we have an opportunity to escape his minions. Without him, they are blind and we can get to Krondor.'

'Maybe,' said Locklear. 'But certainly, if we leave him there, the closer we get to Krondor, the easier it is for his agents to find us.'

'Why?' asked Owyn.

'He's tightening the noose, lad,' said Isaac. 'Less land for his men to cover.'

Locklear said, 'Now Quegans make sense. This Rowe has probably been dealing with Quegan pirates all his life and just sent word to someone in Sarth. First ship outbound to Queg passes word and within a month he's got as many sea-hardened bully-boys as he needs. And if Nago is throwing gold around, there are more Quegans along the roads to Krondor than a beggar has lice.'

'And Quegans aren't likely to run to the King's soldiers if something goes sour; worst they do is skulk back to the nearest port and find a ship heading out. Little chance of being betrayed by someone going cold in the feet,' added Isaac.

'What else?' asked Locklear.

'Nothing,' said Isaac. He stood up and took a cloak off the peg. 'As soon as I pen a note to my cousin, I'm bound for Kesh. I've just set Nago's assassin on my trail, but he doesn't know it yet. Each hour I steal before he does, I stand a better chance of reaching Kesh.'

'I said I'd do you a favour, Isaac, and I will. I'll let you run for Kesh, for old times' sake and for keeping up your end of the bargain, but only if you tell us everything.'

'What makes you think there's anything else?'

Locklear pulled his sword suddenly and had the point at Isaac's throat. 'Because I know you. You always hold something back, just in case you need an edge. I'm guessing this little bit of theatre is to give you a chance to be out of town before us, just in case you can find one of Nago's agents and get him set on us before they figure out you've sold them out. Something like that.'

Isaac grinned. 'Locky! Why I wouldn't – '

Locklear pressed forward with the sword point and Isaac

stopped talking so suddenly he almost swallowed his own tongue. 'All of it,' demanded Locklear in a menacing whisper.

Slowly Isaac raised his hand and gently pushed aside the sword point. 'There's a lockchest – '

'What?' asked Locklear.

Gorath said, 'A chest in which to lock valuables. My people make them to transport items of importance.'

'Go on,' said Locklear.

'There's a lockchest outside of town. Go five miles down the road toward Questor's View. To the right side of the road you'll see a lightning-struck tree. Beyond that is a small clump of brush. Look there and you'll see the chest. I am to leave the ruby there tonight, and when I return tomorrow, my gold is supposed to be waiting for me.'

'So you never see your contact from Krondor?'

'Never. That was part of Nago's instructions to me.'

'You've seen this moredhel?' asked Locklear.

'Met him,' said Isaac. 'At Yellow Mule. He's a big one, like your friend here, not slight like some of them can be. Nasty moods and no humour. Odd fire in his eyes if you know what I mean.'

Locklear said, 'I can imagine. What can you tell us about his company?'

'He only keeps a couple of soldiers around him – I've never seen more than three at any time – because it might be noticed. And there are enough Quegans coming through there that if he needs swords he can get them in a hurry. But he's a magic-user, Locky, a right nasty witch and if you cross him he can fry you as soon as look at you.'

Locklear glanced at Gorath who gave a slight nod of agreement to what was being said. Locklear said, 'Very well, Isaac, here's what you're doing. Get something to write with.'

Isaac glanced around the room and saw an old scrap of faded leather sitting in a corner. He crossed to the small fireplace and fished out some charcoal. He said, 'What do I write?'

'Write this: "Ruby taken by Prince's man. Three you seek are on the way to Eggly. I am undone and must flee." Then sign your name.'

Isaac signed, looking pale as he put down those words. 'This marks me, Locky.'

'You were marked the moment you took gold to turn your hand against your king. You deserve to be hanged, and eventually you will be unless you change your ways, but it will be for another crime, not for this.'

'Unless Nago's agents find you first,' added Gorath.

That was all Isaac needed. 'What do I do with this?'

'Put it in the chest where you are to leave the ruby, then I suggest you start running. If you don't put that note there, and I get to Krondor, I'll hire assassins even if they have to travel to the farthest reaches of Kesh to find you. You can cut your hair and colour it, grow a beard, and wear furs like a Brijainer, but you can't hide that leg, Isaac. Now get out of here.'

Isaac didn't hesitate. He grabbed his sword, his cloak and the note and hurried out the back door.

'How could you spare that traitor?' asked Gorath.

'Dead he is of little use to us, and alive he may direct our foes to another path.' Locklear looked at Gorath. 'And isn't it a little odd you're showing contempt for a traitor?'

The look Gorath returned could only be called murderous. 'I am no traitor. I'm trying to *save* my people, human.' He offered no further embellishment, but turned and said, 'We must be away. That one cannot be trusted and may attempt to bargain for his life.'

Locklear said, 'I know, but either way he plants the note, or he is found and tells them what he knows, which isn't much. They were trying to kill us before we got the ruby. They can't make us any more dead for having it.'

'I think I have a way for us to avoid detection for a while and perhaps reach Nago unseen,' Gorath said.

'How?' asked Locklear.

'I know the way they reach this village of Yellow Mule. If we take the ridge road toward the town you call Eggly, leaving as we told in the note, there's a trail a day's quick run south of here that leads into the higher ridges. It is, I believe, the same trail that empties out near Rowe's farm.'

'How could you know that?' asked Locklear, suddenly suspicious.

Gorath's patience appeared near its end, but he managed to reply evenly. 'Because I lived in these mountains as a child, before you humans came to plague us. Before this land became infested with your kind, my people lived here. I've fished along these rivers and hunted in these mountains.' His voice lowered and he said, 'I may have built my campfire on the spot you humans have built this house. Now, let us go. It's no long journey for a moredhel, but you humans tire easily, and besides, your wounds will slow you even more.'

'And yours won't?' asked Owyn.

'Not so that you would notice,' replied the dark elf, turning to the door without waiting for a response and leaving the building.

Locklear and Owyn hurried after and found Gorath waiting. 'We need to buy food. Have we enough gold?'

'For food, yes,' said Locklear. 'For horses, no.'

They headed to an inn at the east end of town, and Locklear arranged for travel rations, food bound in parchment heavily coated with beeswax, mostly dried or heavily salted to prevent spoilage. While they waited Locklear asked what conditions were like on the road to Eggly, pointedly being loud enough that a few suspicious-looking men hanging about the commons early in the day could overhear. Should anyone ask about them, he was certain this would only reinforce the false information in Isaac's note.

They left the inn and hurried on the road toward the town

of Eggly. Locklear glanced upward, considered the rapidly rising ridge above the trees on the western side of the trail and considered the wisdom of hiking up to that elevation and over the mountains down into a nest of killers over which presided a murderous moredhel sorcerer. Finally he was left with the only answer which he could come up with: there wasn't a better idea presenting itself.

Resigning himself to a long walk and cold nights, he followed Gorath, with Owyn at his side.

FOUR

Passage

The wind howled through the pass.

Locklear spoke through chattering teeth. 'The things I do for king and country.'

Gorath said, 'Ignore the cold. As long as you can feel your fingers and toes, it is only discomfort, nothing more.'

'Easy for you to say,' said Owyn shivering almost uncontrollably. 'You're used to it, living up in the Northlands.'

'You're never "used to it", human. You just learn to accept things over which you have no control.' He looked meaningfully at the two young men, then pointed. 'We can expect to see a sentry any time now.'

'What should we do?' asked Locklear, the cold and his hunger robbing him of his wits.

'Wait over there,' said Gorath, 'while I scout.'

Locklear and Owyn went to the relative shelter provided by the lee side of a huge boulder and waited. Time dragged on and Owyn and Locklear sat close together to preserve warmth.

Suddenly Gorath returned. 'There are four guards near the barn,' he said. 'Within, I do not know, but even alone Nago is dangerous.'

Locklear stood and stomped his feet to restore warmth, flexing fingers and moving in place, getting ready to engage an enemy. 'What do we do?' he asked again, content to let Gorath lead in this circumstance.

Gorath said, 'Owyn, I have no idea of what you are capable, but Nago is a spell-caster of much ability. He can wither a foe with his arts, turning him to lifeless ash, or drive one away screaming in terror. He and his brother are among the most dangerous allies of Delekhan, and serve him even more vigorously since the coming of the Six.'

'Who are the Six?' asked Owyn.

Locklear waved away the question. 'So, how do we deal with Nago?'

Gorath pointed to Owyn. 'You must distract him, boy. Locklear and I will dispatch the other four, and anyone else who might be within the barn, but the magician must be your concern. Cause him to falter, to hesitate, to attempt to leave; anything, but you must keep him there for me to deal with and you must keep him from bringing his arts to bear. Can you do that?'

Owyn was obviously frightened, but he said, 'I will try.'

'No one can ask for more,' said Gorath. To Locklear he said, 'We have surprise, but we must kill the first two quickly. If we are overpowered, or even if we are delayed overmuch in reaching Nago, this will all come to a bad end. If Owyn can't occupy the magician until we reach him, he will end our journey before we can warn your prince.'

Locklear said, 'Then why are we doing this?' Before Gorath could answer, Locklear held up his hand. 'I know, the noose is tightening and if we don't do it now, we will never reach Krondor.'

Gorath nodded. 'Let's go.'

They hurried down the road until they could see the roof of a barn across a small field that sat hard against the ridge. Locklear stooped over, so as to be less visible as they moved down the trail. 'Where are the guards?' he asked Gorath.

'I don't know. They were outside but a moment ago.'

'Perhaps they've gone inside the barn,' suggested Owyn.

Gorath pointed to a notch in the side of the trail, where rain had eroded the soil between two large boulders. He moved between the rocks and slid down the bank to the edge of the field, with Locklear behind and Owyn bringing up the rear.

'We must hurry,' said Gorath. 'The Mothers and Fathers have smiled on us and the guards are inside. We don't know how long this might last.' He set a punishing pace, not wishing to be discovered in the open. Locklear forced himself to push on despite his stiff, aching joints. His wounds had healed, though he still felt weaker than he should. He didn't welcome another fight, but should this Nago be the force behind all the attacks, he welcomed an opportunity to put an end to them, and pay back some of the pain he had been forced to endure.

Gorath reached the barn and huddled in its shadow, glancing in all directions. There was no sign they had been detected. He held up his hand for silence.

They listened. Inside, muffled voices could be heard, though Locklear could make nothing of them, for they were in a tongue he didn't understand. Gorath's hearing was far more acute, for he said, 'They are discussing the fact we have not been seen since Hawk's Hollow. They fear we may have slipped past them on the road through Tannerus.'

'What do we do now?' whispered Owyn.

'As before, we kill them,' said Gorath. 'Act boldly.' He moved to the barn door and withdrew his sword. He pulled forward his hood, throwing his features into darkness, then put his sword under his cloak and turned to Owyn and Locklear. 'Be ready, but wait a moment before entering.'

Then Gorath pushed open the door and in the late-afternoon gloom must have seemed a black shape against a darkening sky. From within a voice sounded a note of inquiry. Gorath stepped forward with a stride that communicated purpose, answering in the moredhel tongue. He must have confused them for a moment, for one asked another question before a different voice shouted, 'Gorath!'

Locklear didn't hesitate when he heard that, but virtually jumped through the open door. Owyn was a step behind.

The barn was empty save for five moredhel. A table had been placed in the centre of a large barn aisle, with a bench behind it, where the moredhel magician Nago was rising in shock at the appearance of his intended prey.

A moredhel guard was falling from Gorath's first blow as he rounded on another, lashing out with his blade and forcing the swordsman backward, clutching his bleeding sword arm. Locklear dashed forward and caught the wounded dark elf from behind, killing him with a blow to the back of his neck as he sought to disengage himself from Gorath's attack, leaving both swordsmen facing a ready opponent.

Owyn saw the moredhel magic-user who was still motionless in astonishment at the appearance of the prey he had been seeking for weeks. But as Owyn moved through the doorway, he felt power beginning to manifest as Nago started an incantation. Knowing there was nothing much he could do, Owyn unleashed the only spell he could throw on short notice, the blinding spell he had practised so much on the journey.

The dark elf blinked in surprise and faltered, breaking his spell. Owyn hesitated then raised his staff and started his charge, doing his best to imitate a warcry. A thin warbling sound escaped his lips as he ran between Gorath and Locklear as they struggled with their opponents.

As he closed upon the moredhel magician, Owyn slipped and fell forward, which saved his life, for the enraged Nago unleashed a bolt of shimmering purple-and-grey energy which sped through the spot where Owyn had been a moment earlier. Rather than strike the lad full on, it brushed over his back, and where it touched Owyn felt agony, a shocking pain. His head swam from it, and he felt dizzy. The muscles in his lower back and legs refused to obey him. He struggled, but they felt encased in metal bonds.

Rolling over, Owyn saw the magician begin another spell, and without any other option, Owyn threw his staff at the moredhel. As he expected, the magician ducked aside, and his spell-casting was interrupted. Nago shut his eyes, as

if in pain, and Owyn knew the enemy spell-caster was
struggling to restart his spell. While only a novice at magic,
Owyn understood enough of it to know that an interrupted
spell could prove painful and that it might take Nago a few
moments to refocus his thoughts and regain the ability to
inflict harm upon his opponent.

Owyn tried to focus his own thoughts, as if he might
throw another spell to distract Nago a moment longer, but
his own thinking was chaotic, his mind racing with conflict-
ing images. Phrases and concepts previously unknown to
him intruded into his concentration and he couldn't force
himself to come up with any useful conjuration. He fumbled
in his belt for a dagger and thought to throw that at Nago.

Nago opened his eyes and looked past Owyn, to where
the struggle was ending. Owyn rolled over and saw Gorath
running his opponent through, while Locklear seemed to be
getting the best of his own. Owyn looked over his shoulder
at Nago and saw the magician was hesitating, then starting
to turn to flee.

'He's trying to escape!' Owyn shouted, but his voice was
weak and he didn't know if he had warned his companions.

Gorath heard and was past Owyn in three huge strides.
The moredhel magician turned and threw something at
Gorath, and sparking energies coursed around the dark elf
chieftain. Gorath groaned in pain and faltered.

Owyn threw his dagger, a weak underhand cast, but one
which caused the butt of the weapon to strike Nago in the
temple. As if released from a prison, Gorath rose up and with
a single blow struck Nago in the neck, nearly severing his
head from his body.

Locklear hurried over and helped Owyn to his feet. 'We
could have used a prisoner,' he observed.

Gorath said, 'These guards know nothing worth learning.
And Nago could not be left alive. While you were trying to
question him, he would have been sending word to his
confederates that we are here.' The dark elf looked down

at Owyn who still lay on the floor. 'You did well, boy. Are you all right?'

'My legs don't work,' he answered. 'I think I will get them to work in a while.'

'I hope so,' said Locklear. 'I'd hate to leave you here.'

'I'd hate to be left,' said Owyn.

Gorath looked around. He moved to a large cache of provisions and dug out some bread and a waterskin. He took a drink, handed it to Locklear and tore the loaf into three portions, handing one each to the other two.

Locklear helped Owyn sit up at a table and looked at a map unrolled there. What have we here? he asked himself as he studied the map.

It was a map of the area south of Hawk's Hollow, with guard locations marked and fresh ink indicating sightings. It was clear that they had avoided detection from Hawk's Hollow to Yellow Mule. Locklear said, 'Owyn, could Nago have got word out to others that we are here?'

Owyn felt his legs with his hands as if trying to determine what was wrong with them and said, 'It's doubtful. I kept him busy and he was trying to kill us. I can imagine he could do two things at once, but three is unlikely. If he's got a routine for checking in with his agents, they'll soon know something is wrong because of his *not* contacting them.'

'Then we must be on our way,' said Gorath. 'How far to Krondor?'

'If we were taking a stroll down the King's Highway without fear, another two days. By horse, less than a day from here. Through the woods, maybe three.'

Gorath asked Owyn, 'How long before you can move?'

'I don't know – ' Then suddenly Owyn's legs moved. 'I guess I can move now,' he said, rising slowly. 'Interesting,' he said.

'What's interesting?' asked Locklear.

'That spell. It's designed to bind an opponent, but only for a short while.'

'Why is that interesting?'

'It's some sort of combat magic. They don't teach that at Stardock.'

'Can you do the same thing?' asked Gorath. 'It could prove useful.'

'Really?' asked Locklear dryly.

'I don't know,' said Owyn. 'When the spell struck me, something happened, a recognition of some sort. I will think on it, and maybe I can figure out how he did it.'

'Well, figure out how while we're moving, assuming you're ready to go,' said Locklear around a mouthful of bread. They quickly rummaged through the cache of supplies and found several dark grey-blue fur-lined cloaks. 'These will serve us well,' said Locklear, still warm from the fight, but knowing all too well how cold the nights were along the coast this time of the year. Locklear gathered up the maps and several messages, all claiming forces were in place for key attacks at various locations throughout the west. He placed those in a pouch and slung it over his shoulder.

They left the barn and circled around the darkened farmhouse. The owner was either sleeping or dead, betrayed by his guests, but either way they did not wish to spend time finding out. They had three dangerous days before them and knew there were perils enough along the route to Krondor without stopping to look for them.

Twice they had avoided assassins or bandits; they didn't know which. Once they had lain in the mud in a gully next to a woodland path while a band of armed Quegans had hurried past. Now they stood behind the last line of trees before open farmland. Beyond they could see the City of Krondor.

'Impressive,' said Gorath in a neutral tone.

'I've seen Armengar,' said Locklear. 'I am surprised to hear you call this impressive.'

'It's not the size of the place,' said Gorath. 'It's the hive of humans within.' For a moment he looked off into the distance. 'You shortlived creatures have no sense of history or your place in this world,' he said. 'You breed like – ' He glanced over to see Locklear's dark expression and said, 'No matter. There are just a great deal of you at any one time in any one place, it seems, and this is more of you in such a small place.' He shook his head. 'For my people, such gatherings are alien.'

'Yet you rallied at Sar-Sargoth,' observed Locklear.

'Yes we did,' said Gorath. 'To the sorrow of many of us.'

Owyn said, 'Do we just walk across this field to the road?'

Locklear said, 'No. Look over there.' He pointed to a place where a small farm road intersected the King's Highway. A half-dozen men stood idly by as if waiting for something. 'Not exactly a place to hoist a few and talk of the day's labours, is it?'

'No,' said Owyn. 'Where do we go then?'

'Follow me,' said Locklear as he moved along the tree line, farther east. They reached a long gully, a naturally occurring watercourse that would be flooded when the thaw came to the mountains to the north and east, but which currently hosted only a small stream. 'This runs to a place by the eastern gate, in the foulbourgh.'

'Foulbourgh?' asked Gorath.

'The part of the city built outside the wall. There are ways to get in and out of the city if you know them. The sewers under the foulbourgh and city proper are not supposed to connect, so an enemy can't use them to gain entrance.'

'But they do,' supplied Gorath.

'Yes, in two places, and one of them is as dangerous as walking up to those men gathered back there and asking for directions to the Prince's palace. That entrance is controlled by the Thieves' Guild. But the other entrance – well, let's say that besides a friend of mine, only a few others know of it.'

'How is it you know of it?' asked Gorath.

'My friend and I used it once, a long time ago, to follow Arutha to Lorien.'

Gorath nodded. 'We have heard of that encounter. Murmandamus's trap to kill the Lord of the West.'

'That's the one,' said Locklear. 'Now, it would be a good time to move silently.'

They did as Locklear bid and moved through the gully, until they encountered a culvert, made of stones polished by the water over the years. They bent over and walked below the road, as the late-afternoon shadows lengthened. Finally, the culvert ducked under a small stone bridge that afforded them a hiding place. It was well shielded from prying eyes by stores stacked in crates on each side of the road waiting for transport. Bored workers slowly moved to load them.

'We linger a bit, until it gets darker,' said Locklear. 'At the right time, we need to get up and blend in with some traffic heading along the road that runs beside this culvert.' He went to the other side of the bridge and glanced upward, pulling his head back.

Pointing where he had looked, he said, 'Someone's hanging around up there.'

'What do we do?' asked Gorath, obviously as out of his element as Locklear had been on the mountain trail.

'We wait,' said Locklear. 'A patrol from the city watch passes along here about sundown, and they'll order any armed men to move along. After dark it gets dangerous outside the wall, and the watch doesn't like too many swords gathered in one place.'

They sat under the bridge, in the puddles on either side of the stream, waiting in silence as the hours dragged by. Flies annoyed them, and only Gorath ignored their presence as Locklear and Owyn spent most of the time swatting them away.

As sundown approached, Locklear heard the tread of boots upon the cobbles above. A few voices were raised, and Locklear said, 'Now!'

He moved quickly up the side of the bank just beyond the bridge, ducking behind some crates as a party of men dispersed under the watchful eye of the city guard. 'They'll come this way, back toward the palace,' said Locklear. 'We just duck in beside them, and even if we're seen, it's unlikely we're going to be attacked with a dozen soldiers ready to start busting heads at the first sign of trouble.' He pointed to Gorath. 'But you'd better fix that hood. Most people here wouldn't know an elf from a moredhel if you hung signs around your neck, but you never know. If Ruthia's fickle, the first person we meet will be an old vet from the wars to the north.' Ruthia was the Goddess of Luck.

Gorath did as he was told and pulled his hood forward, hiding his features and when the soldiers walked down the road beside the stream, he followed Locklear and Owyn as they hurried to match pace with the soldiers.

They walked from the northeasternmost corner of the city along its entire length to the southern gate, and when the city watch moved toward the palace entrance, Locklear pulled them aside.

Owyn said, 'Why don't we just follow them in?'

'Look,' said Locklear. They looked where he pointed and saw a work crew gathered before the gate, with two teams of horses tied to a pulley. 'It seems someone has sabotaged the gate,' said Locklear.

The watch commander shouted something down from the wall to the patrol leader, who saluted and turned his men around. 'Come on, lads,' he said, 'we're for the northern gate.'

Locklear motioned for his companions to follow him and he led them through a back alley. 'This way,' he urged.

He took them to what appeared to be the back entrance to a small inn, and opened the gate. Once through, he closed the gate and they stood in a tiny stabling yard, with a small shed off to one side. Looking to see if they were observed, Locklear pointed to the rear door of the inn. 'If anyone finds us, we're lost, looking for a meal and once we get inside the inn, head toward the front door; if anyone objects, we run like hell.'

Gorath said, 'Where are we?'

'The back of an inn owned by people who would be less than pleased to discover we knew about this place, or what I'm about to do.' He moved toward the shed, but rather than going inside, he moved to where it joined with the wall. Feeling around behind the shed, Locklear tripped a lever and a latch clicked. A big stone rolled away, and Owyn and Gorath could see it was a cleverly-fashioned sham, made of canvas and painted to look like the rock of the wall. Locklear was forced to lie down and wiggle feet first through the small aperture, but he successfully negotiated the entrance. Owyn went next, and Gorath last, barely clearing the opening.

'Who uses that thing?' asked Owyn in a whisper. 'Children?'

'Yes,' said Locklear. 'The Mockers number many urchins in their ranks and there are dozens of bolt-holes like that all over the city.'

'Where are we?' asked Owyn.

'Use your senses, human,' said Gorath. 'Or can't your breed smell its own stink?'

'Oh,' Owyn exclaimed, as the stench of the sewer struck him.

Locklear reached up and pulled shut the trap, leaving them in total darkness.

'My kind see in darkness better than yours do, Locklear,' said Gorath, 'but even we must have some light.'

'There should be a lantern close by,' said Locklear. 'If I can remember the distance . . . and direction.'

'What?' asked Gorath. 'You don't know where a light is?'

'I can help,' said Owyn. A moment later a faint nimbus of light started to glow around the young man's hand, and it grew until they could see a dozen paces in all directions.

'How did you do that?' asked Locklear.

Owyn held out his left hand. On it was a ring. 'I took it off Nago. It's magic.'

'Which way?' asked Gorath.

'This way,' said Locklear, leading them into the sewers of Krondor.

'Where are we?' whispered Owyn.

Locklear lost his sure tone as he said, 'I think we're just north of the palace.'

'You think?' said Gorath with a snort of contempt.

'All right,' said Locklear with a petulant tone. 'So I'm a little lost. I'll find – '

'Your death, quick and messy,' said a voice from outside the range of Owyn's light.

Three swords cleared their scabbards as Locklear tried to pierce the gloom beyond the light by force of will.

'Who be you and what would you in the Thieves' Highway?'

Locklear cocked his head at the bad attempt at a formal challenge and, judging the owner of the voice to be a youth, he answered, 'I be Seigneur Locklear and I do whatever I will in the Prince's sewers. If you're half as intelligent as you're trying to sound, you'll know not to bar our way.'

A young boy stepped forward from the shadows, slender and wearing a tunic too large for him, wrapped around the waist with a rope belt, trousers he had almost outgrown, and sporting a pointed felt cap. He carried a short sword. 'I'm Limm and fast with a blade. Step any further without my leave and your blood will flow.'

Gorath said, 'The only thing you'll do is die, boy, if you don't stand aside.'

If the towering presence of the moredhel chieftain had any effect on the lad, he hid it as he bravely said, 'I've bested better than you when I was a boy.' He stepped back, cautiously. 'And besides, I've got five bashers back there waiting for my call.'

Locklear held up his hand to restrain Gorath. 'You remind me of a young Jimmy the Hand,' said Locklear. 'Full of bluster as well as guile. Run off and there's no need for anyone's blood to flow.' Softly to Gorath he said, 'If he has bashers nearby, we don't need the trouble.'

'Jimmy the Hand, is it?' asked Limm. 'Well, if you're friends of Seigneur James, we'll let you pass. But when you see him, tell him he had better come soon or the deal is off.' Before Locklear could answer, Limm was deep in shadows, so silently they could barely hear him move. From a distance he said, 'And watch your step, Locklear who knows Jimmy the Hand. There are nasty customers nearby.' As the voice faded, Limm added, 'And you're completely turned around. Turn to the right at the next culvert, and straight on until you reach the palace.'

Locklear waited, listening for more. But only silence punctuated by the trickling sound of water and the occasional echo of some distant sound in the sewer could be heard.

Gorath said, 'That was passing strange.'

'Yes,' agreed Owyn.

'More than you know,' said Locklear. 'That boy was waiting for my friend James. And James has the death mark on him from the Mockers if he ever trespasses their territory. That was a deal struck by Prince Arutha for James's life years ago.'

Owyn said, 'Sometimes agreements change.'

'Or are broken,' added Gorath.

Locklear said, 'Well, we'll sort this out later. Right now we need to find our way to the palace.'

'What did he mean by "nasty customers nearby"?' asked Owyn.

'I don't know,' answered Locklear. 'I have a feeling if we're not careful we'll find out,' he whispered.

They turned in the direction instructed by Limm and

moved to the corner where he had told them to turn. A short way along the indicated route, Gorath said, 'Someone ahead.'

Owyn put his ring under his arm, causing the light to diminish. 'Two men,' whispered Gorath. 'Wearing black.'

'Which is why I can't see them,' said Locklear.

'Who are they?' asked Owyn.

Locklear turned and knew his withering look was lost in the gloom, so he said, 'Why don't you just go up and ask them.'

'If they aren't the Prince's men or those Mockers, then they must be enemies,' said Gorath, stepping forward quickly, his sword ready to deliver a killing blow.

Locklear hesitated a moment, and by the time he started moving, the dark elf was upon the two men. The first turned just in time to see his own death arrive, for Gorath slashed him deeply across the chest and shoulder.

The second man drew his sword and attempted to slash down on Gorath's head, but Locklear stepped in and parried the blow high, allowing Gorath to run him through. It was over in seconds.

Locklear knelt and examined the two bodies. They wore identical trousers and tunics of black material, and black leather boots. Both men had short swords and one had laid aside a short bow within easy reach. Both men were without purse or pouch, but both wore identical medallions under their tunics.

'Nighthawks!' said Locklear.

'Assassins?' asked Owyn.

'But they should have . . .' Locklear shook his head. 'If these two are Nighthawks, I'm Gorath's grandfather.'

Gorath snorted at the idea, but said, 'We have heard of your Nighthawks; some were employed by agents of Murmandamus.'

Owyn said, 'The stories are they had nearly magical abilities.'

'Stories,' said Locklear. 'My friend James faced one on the rooftops of the city when he was no more than a lad of fourteen years and lived to tell the tale.' Locklear stood. 'They were good, but no more than other men. But the legend helped them get their price. But these,' he indicated the two dead men, 'were not Nighthawks.'

A whistle sounded from down a nearby tunnel. Gorath spun, his sword ready to face another attack. Locklear, however, just put two fingers to his mouth and whistled in return. A moment later a young man stepped into the light. 'Locky?' he asked.

'Jimmy!' said Locklear as he embraced his old friend. 'We were just speaking of you.'

James, squire of the Prince's court, regarded his best friend. He took in the long hair gathered behind in a knot and the bushy moustache and said, 'What have you done to your hair?'

'I haven't seen you in months and the first thing you ask about is fashion?' asked Locklear.

James grinned. His face was youthful, though he was no longer a boy. He had curly brown hair he kept cropped short and was dressed in plain clothing, tunic, trousers, boots and cloak. He carried only a belt knife. 'What brings you back to court? Arutha banished you for a year, if memory serves.'

'This moredhel,' said Locklear. 'His name is Gorath and he brings a warning to Arutha.' Pointing to his other companion, he said, 'And this is Owyn, son of the Baron of Timons. He's been of great help to me, also.'

James said, 'A moredhel chieftain in Krondor. Well, things are getting strange hereabouts, too.' He glanced down at the two dead men. 'Someone has bribed a few very stupid men to play the part of Nighthawks, here in the sewers and in other parts of town.'

'Why?' asked Locklear.

'We don't know,' said James. 'I'm on my way to meet with some . . . old acquaintances of mine. To see if we can cooperate in uncovering who is behind this mummery.'

'The Mockers,' said Locklear. 'We ran into one of them, a lad named Limm.'

James nodded. 'I'm to meet with some of them shortly. I had better not disappoint them. But before I go, what are you doing down here in the sewer?'

Locklear said, 'Someone wants Gorath dead very badly. I've been cut more times than a horse's flank by a cheap butcher. We're here because we need to get into the palace, and I've seen lots of very dangerous-looking men watching the entrances to the palace. When I tried to get us in by shadowing the city watch trying to enter, we found the gate damaged.'

'Someone sabotaged it, as well as the north palace entry. The only way into the palace right now is through the sea-dock gate, or here.'

Locklear looked concerned. 'They even had the gate jammed to keep us from reaching the palace?'

James nodded. 'That would explain the mystery. Look, go see Arutha and I'll catch up with you later.'

'That's the way?' asked Locklear.

'Yes,' said James. He fished out a key and handed it to Locklear. 'But we've locked the secret door so you'd have had a long wait if I hadn't chanced by.'

'I might have picked the lock,' said Locklear. 'I've watched you do it a few times.'

'And pigs might fly,' said James with a pat on his friend's shoulder. 'It's good to see you back, even if under such dark clouds.' He pointed the way he had come. 'Make your way past two large culverts on the left, and you'll find the ladder to the palace.' With a departing grin, he added, 'I suggest you bathe before calling upon Arutha.'

Locklear smiled, then laughed. For the first time in months he suddenly felt safe. They were but a short walk away from the entrance to the palace, and he knew that soon he would be enjoying a hot bath. 'Come see me when you've returned,' he said to James. 'We have much to catch up on.'

'I will,' said James.

Locklear led Gorath and Owyn to the ladder that led up into the palace, a series of iron bars hammered into the stones rising a floor above. There a grate with a heavy lock had been erected, and Locklear used the key James had provided to open it. They swung aside the grate and moved into a small tunnel just above the sewers, leading into the lower basement of the palace. Locklear silently led them to a door. Once through, Owyn and Gorath saw they were in another passage, this one lit by torches in widely separated sconces, and when the door was returned to its resting place, it vanished into the stone wall.

Locklear led them to his quarters, past a pair of palace guards who only watched with interest as the Prince's squire walked past with another youth and what looked like a tall elf.

Glancing through a window overlooking the city, Locklear said, 'Suppertime's in about an hour. Time for a bath and a change of clothing. We can talk to the Prince after the meal.'

Gorath said, 'It seems so . . . odd to be here.'

Opening the door to his quarters, Locklear said, 'Not nearly as odd as having you here.' He stepped aside to admit his guests, and turned to wave at a page hurrying down a nearby hall. 'Boy!' he shouted.

The page stopped and turned to run toward him. 'Sir?' he said.

'Send word to the Prince that I've returned with a message of the gravest consequence.'

The boy, who knew Locklear well, indulged himself in an observation: 'It'll be grave, all right; your grave, if the Prince doesn't agree, squire.'

With a playful slap to the side of the head, Locklear

sent him off. 'And pass word I need enough hot water for three baths!'

The boy waved he had heard and said, 'I'll tell the staff, squire.'

Locklear turned into his room and found Owyn sitting on his bed, lying back against the wall. Gorath stood a short way off, patiently waiting. Locklear went to his wardrobe and selected some clothing. 'We'll send for something closer your size while we bathe,' he said to Gorath. He took the clothing and handed a tunic and trousers to Owyn, along with fresh smallclothes, then said, 'This way to the bath, my friends.'

At the end of the hall he found four servants pouring hot water into a large tub, while another waited. 'In you go,' he said to Owyn, who stripped off his filthy garments and climbed into the tub. He settled in with a satisfied 'ah' sound and rested back in the hot water.

Gorath said, 'Is that third tub for me?'

'I was going to take that one, but if you – '

'Fill it with cold water.'

The servants exchanged glances, but Locklear nodded, so they finished filling the second tub and ran off, turning around a pair of servants hurrying from the kitchen with hot buckets. Soon they returned with cold water and started filling the tub.

Gorath stripped and climbed in, allowing them to pour the cold water over his head. He endured the cold water without comment. When they were done bathing and clean clothes had been fetched for Gorath, Owyn asked, 'Why cold water?'

'We bathe in mountain streams in a land that always sees ice upon the peaks,' said Gorath. 'This water was too warm for my taste.'

Locklear shrugged. 'You learn something new every day.'

'Yes,' agreed Gorath. 'You do.'

When they were dressed, they left the bathing chamber to

discover a squad of palace guards waiting for them. 'We're to escort you to the Prince, squire.'

Locklear dryly said, 'No need. I know the way.'

The sergeant, a tough old veteran, ignored the young noble's marginal rank and said, 'The Prince thought there was a need, sir.'

He signalled and two soldiers fell in on either side of Gorath and two fell in behind him. They moved along the hall until they were ushered into the dining hall, where Prince Arutha, Princess Anita and their guests were finishing their dinner.

Arutha, ruler of the Western Realm of the Kingdom of the Isles, sat at the centre of the head table. He was still a young man. Despite having ruled the realm for ten years, his face was only now starting to show the lines which age and responsibility bring. He kept his chin shaved, so that he still resembled the youth who had emerged a hero of the Riftwar. His hair was mostly black with a few stray grey hairs beginning to show, but otherwise he looked much as he had when Locklear had first come to Krondor, a page boy fresh from his father's court at Land's End. His brown eyes settled on Locklear with a gaze that had reduced lesser men to trembling children over the years; Locklear had endured that gaze many times in the ten years he had served in Arutha's court.

Princess Anita favoured Locklear with a smile, her green eyes alight at one of her favourite courtiers returning after a long absence. Locklear, like the other younger men in the court, almost worshipped the Princess for her effortless grace and genuine charm.

At the table were others known to Locklear: Gardan, Knight-Marshal of the Principality; Duke Brendan, Lord of the Southern Marches; and others. But near the Princess's seat was one who was unknown to Locklear; a man wearing the black robe of a Tsurani Great One. He had receding snow-white hair that fell to his shoulders. His eyes fastened upon Locklear, and Owyn could sense that this was a man who possessed powers rivalled by few in the world. Locklear knew it must be Makala, the Tsurani Great One come recently to this court.

'Seigneur,' began Arutha, formally, 'you were ordered to attend to the needs of the Earl of Tyr-Sog for a year. By my calculations, you are many months short of that duty. Have you a persuasive reason for ignoring my orders?'

Locklear bowed and said, 'Highness. Only the most grave tidings from the north would have me quit my post and hasten here. This is Gorath, Chieftain of the Ardanien, who has come to warn you.'

'Warn me of what, moredhel?' asked Arutha with a suspicious gaze. His previous experience with the moredhel was murder and deception.

Gorath stepped forward. 'I warn you of war and bloodshed. The war drums beat at Sar-Sargoth once more and the clans gather.'

'For what purpose?' asked Arutha.

'Delekhan, Chieftain of the Darkanien, gathers the clans. He sings songs of power and musters to return south.'

Arutha said, 'Why? For what purpose?'

Gorath said, 'He swears that Murmandamus lives, and that you hold him captive in the city of Sethanon. And he swears by the blood of our ancestors we must return to free our leader.'

Arutha sat stunned. He had killed Murmandamus, though few had witnessed the duel. He also knew that Murmandamus had been a fraud, perpetrated by the Pantathian Serpent Priests to gull the moredhel into serving their dark cause.

Arutha stood. 'We will speak of this in my private council.' He bowed to his wife, then motioned to Makala. 'If you would join us?'

The Tsurani magician nodded and rose, and Locklear saw he was unusually tall for a Tsurani, perhaps five feet ten inches in height. Makala spoke briefly to a servant, who bowed low and hurried off to do his master's bidding.

Locklear motioned for Owyn and Gorath to accompany him through large doors on the right of the dining hall, the entrance into the Royal Family's private apartments. To Gorath he said, 'I hope you have more to tell Arutha than that, or we're both in deep trouble.'

'More trouble than you know, human,' said Gorath.

FIVE

Mission

Drums thundered across the ridges.

Gorath stood rooted in confusion. Part of him knew this was a memory, yet the experience was as real as when he had lived it. He clutched his hands and looked at them. They were small, a child's hands. He glanced down and saw bare feet, and he had not gone barefoot since he was a boy.

Atop the surrounding hills drummers pounded out their insistent rhythms as fires burned brightly in the night. Clans long at war with one another watched for signs of betrayal, but all had come to hear the Speaker. Gorath stumbled along, his feet leaden with mystic fatigue; no matter how hard he tried, he could not move quickly.

The peace had fractured; he knew this. He knew his father's people had been betrayed. He was but twelve summers of age and it should be centuries before the mantle of leadership fell to him, but fate ruled otherwise. Without being told he knew his father was dead.

His mother came up behind him and said, 'Move quickly. If you are to lead, you must first survive.' Her voice echoed and was distant and when he turned to look back at her, she was gone.

Suddenly he stood dressed in armour and boots; too big for him yet they were his own. His father had fallen when the Speaker's peace had dissolved in fury. Like others before him, the Speaker had sought to raise the banner of Murmandamus, the only leader ever to unite the numerous clans of the moredhel. Now Gorath, a boy barely able to hold his dead father's sword, stood before the men of the Hawk

Clan, as dispirited a lot as had ever gathered around the fire. Gorath's mother tapped him on the shoulder and he turned. 'You must say something,' she whispered.

Looking at the men of his clan, Gorath could barely make a sound, yet these warriors, some alive more than a century, waited to hear a boy's words. The words that were to lift them from the depths of their hopelessness. Looking from face to face, at last Gorath said, 'We will endure.'

A wave of pain gripped Gorath and he fell to his knees, and suddenly he was a man, kneeling before Bardol, swearing alliance in exchange for protection. Bardol had no sons and needed a strong husband for his daughter. Gorath had proven himself a wily leader, taking his people high up into the great ice mountains, living in caves lined with lichen, hunting bear and reindeer. For twenty-five years his people had survived, healed, and when he returned home, he had hunted down his father's betrayer. He had entered the camp of Jodwah and thrown down the head of his brother, Ashantuk, at his feet in defiance. Then he had killed Jodwah in fair combat, and the warriors of the Lahuta, the Eagle Clan of the Northern Lakes had joined with the Hawk Clan of the Ice Peaks, and Gorath had emerged the leader of the Ardanien, the flying hunters in the ancient tongue. And he was but a stripling of thirty-seven summers, yet he commanded more than a hundred warriors.

Twice more he had come to council called by chieftains who had claimed rights beyond their reach, and he had watched as battles had bled his people. He had been clever and kept his people outside such conflicts, and he had become a man to be sought out, to give counsel, because he had no ambitions of his own. Many trusted Gorath. He was approaching his prime and numbered a hundred and six years of age. A thousand swords did his bidding.

Time was a river, and he swam in it. Wives – two women who had borne him children – he had seen the first dead from a human arrow: the other had left him. He had sons and a daughter, though none alive now. For even Gorath, he who was trusted for his wise counsel and cautious ways, even he had been swept up in the madness that had been Murmandamus.

The one called Murmandamus had returned, as spoken of in the prophecies. He wore the mark of the dragon and possessed great powers. He was served by a priest of a far people, a creature who hid in heavy robes, and first among his followers was Murad, Chieftain of Clan Badger of the Teeth of the World. Gorath had seen Murad break a warrior's back over his knee and knew that only the most powerful leader could command Murad's allegiance. As a sign of Murmandamus's potency, Murad had cut his own tongue, proof he would never betray his master.

For the only time in his life, Gorath was caught up in madness. The blood pounded in his ears in harmony with the thunder of war drums in the mountain. He had led his army to the edge of the great Edder, and had fought the mad ones, Old King Redtree's barbarians, and had held the flank while Murmandamus assaulted the human city of Sar-Isbandia, what the humans called Armengar.

Thousands had died at Armengar, but his clan was whole. A few had fallen holding the flank against the forest and on the march through the pass the humans called Highcastle. There, at Highcastle, he had lost Melos, his blood kin, son to his mother's sister. There at Highcastle, a third of the Ardanien had perished.

Then had come Sethanon. The fighting had been brutal, but the city had been theirs. Yet at the moment of triumph, victory had been taken from them. Murmandamus had vanished. According to some of the warriors one moment he had stood in the barbican of the castle at Sethanon, and the next he was gone. Then the Keshians had arrived, and the Tsurani, and the battle had turned. The giants recruited from their high villages had been the first to flee, then the goblins, courageous when victorious, but quick to panic, had left the battle. It had been Gorath, the only surviving chieftain at the castle who had been the first to call the withdrawal. He had come looking for the master, because fighting had erupted between two rival clans over spoils, and only Murmandamus could settle the dispute. Humans had escaped because of the fighting. No one could find the master, and Gorath had cursed all omens, prophecies and heralds of destruction, and had returned to gather the Ardanien and lead them northward.

Most of his warriors had survived, but many chieftains labelled Gorath and his followers as betrayers. For nine summers, the Ardanien lived in their valley, high up in the northern mountains, keeping their own counsel. Then had come the call.

The banners were again raised and it was Delekhan, sworn enemy – son of the man who had slain Gorath's father, and who had died at Gorath's hands in turn – blood enemy from birth, who rallied the clans. Delekhan who had eaten with Murad and the snake priest, and who had been the last surviving member of Murmandamus's council. And it was Delekhan who vowed that Murmandamus still lived within a prison in the heart of Sethanon and only by freeing him could the Nations of the North take back the land seized by the hated humans.

And any who spoke against Delekhan was struck down. Dark magics were fashioned by the Six, and one by one the opponents of Delekhan's plan vanished. Gorath knew his day was coming, and knew that he must carry word to his enemies to the south, for they were his people's only hope.

Night, and he fled through ice and pain. Men who were once as brothers to him sought to hunt him down and end his life. Haseth, whom Gorath had taught to hold a sword, last among his blood kin, had led them. It had been by Gorath's own hands that his last surviving kinsman had died.

Then again, he heard the thundering drums. Again he saw

the fires on the hill, but now he felt his mind returning to the present, memories of his life fading away slowly . . .

The girl was young, not quite seventeen years of age, yet her hair was nearly white with only the faintest hint of gold in it. Pale eyes of blue regarded Gorath as she let go of his hands. Behind her stood the Prince of Krondor, the black-robed Tsurani, and another spell-caster, one who, while short of stature, was almost exuding power. Others were nearby, but those Gorath had travelled with, Owyn and Locklear, were in another room.

'What did you see?' asked the Prince.

'I cannot find any falsehoods, Highness,' said the girl in a weary tone. 'But I cannot find the truth, either. His mind is . . . alien, chaotic.'

Prince Arutha's brown eyes narrowed as he regarded Gorath. 'He hides his thoughts?'

The bearded magician said, 'Highness, Gorath is moredhel, and even with Gamina's exceptional talents for reading thoughts, his mind may have many innate psychic defences. We have never had the privilege of studying a moredhel. From what I learned in my time with the eldar – '

At mention of the ancient elven lore keepers, Gorath's eyes narrowed. 'You are Pug,' he said.

Pug nodded. 'I am.'

'We have heard of you, who studied with the eldar,' said Gorath.

Arutha said, 'The point?'

'I think he's telling the truth,' said Pug.

'As do I,' said Makala. 'Forgive me,' said the Tsurani magician to Prince Arutha, 'but I presumed to use my own arts to watch as the Lady Gamina examined the moredhel. It is as she has stated; there is confusion and an alien mind there, but no guile. Despite his differences from us, he is as honest a creature as you will meet.'

'For what cause did you presume to use your arts without

leave?' asked Arutha. His tone was one of pointed curiosity, rather than anger.

'War in the Kingdom would have many wide-ranging consequences, not the least of which would be a disruption of trade between our two worlds, Your Highness. The Light of Heaven would be most displeased if such occurred, let alone the risk if such as these – ' he indicated Gorath ' – gleaned the secrets of the rift.'

Arutha nodded, his expression thoughtful. Gorath spoke. 'Trading agreements notwithstanding, war benefits no one, Prince. Despite that, you must prepare your army for war.'

Arutha's words were pointed, but his tone was even. 'What I must or must not do will be my burden, renegade. And my decisions will be based upon more than simply the word of one dissident chieftain. If not for Locklear's faith in you, you'd be in our dungeon making the acquaintance of our torturer, not holding hands with Lady Gamina.'

Gorath glared at the Prince of Krondor. 'I would tell you no different under hot iron, the lash, or the blade, human!'

Pug asked, 'Then why do you betray your own, Gorath? Why come to Krondor with a warning when your nations have sought to dislodge humankind from this world as long as either race can remember? Why betray Delekhan to the Kingdom of the Isles? Are you seeking to have our army do what you cannot do by your own might, and destroy an enemy?'

The dark elf studied the magician. Despite his youthful appearance he was a man of great power, and to this point he had spoken to Gorath only in tones and terms of respect. Softly, Gorath said, 'Delekhan may be a bitter draught to the Kingdom, but he is poison in the throats of our people. He enslaves and conquers, and he seeks to claim greatness, but – ' He took a deep breath.

'My people are few in number,' he said slowly. 'We will never count as many swords and arrows as you humans. We rely upon those who willingly serve us, the goblins, mountain giants, trolls, and renegade men.' His tone took on a bitter edge. 'Two sons and a daughter I have mourned, and of two wives, one I have seen travel to join the Mothers and Fathers, while the other left me for being the one to call retreat at Sethanon. My last blood kin died at my own hands the night I met young Owyn.' He looked directly into Arutha's eyes. 'I can never go back, Prince of Krondor. I will die in an alien land among people who despise my race.'

'Then why?' asked Arutha.

'Because my people cannot withstand another war such as we had at Sethanon. Delekhan appears, wearing the dragon helm of Murmandamus, and swords are raised and blood oaths sworn, but while we have courage and dedication in abundance, we lack strength of numbers. Should enough of us die in futility again, the Northlands would lie open to human conquest. We would be as echoes on the wind, for within a hundred years no moredhel would remain alive.'

'We are content to stay on this side of the Teeth of the World. We have no ambitions in the Northlands,' said Arutha.

'You may not, here in your warm castle in Krondor, Prince, but there are those among your race would conquer to win a title, and you know this. If one came to your King with word that he had seized the town of Raglam and had occupied Harlik, and now controlled a third of the Northlands, would your King offer him a hereditary title and income from those lands?'

'He would,' admitted Arutha.

'Then you see my point,' said Gorath.

Arutha rubbed his chin. He stood lost in thought a long moment, then said, 'You are persuasive, Gorath. I will take what Makala and Gamina say at face value and assume you have no guile in you. But what now must be decided is if what you know to be true is, indeed, truth.'

'What do you mean?' demanded Gorath.

Pug said, 'What he means is you may be an unwitting tool. If this Delekhan knew of your animosity, might he not have given you the information you seek to bring to us, to cause us to rush to meet him at some place of his choosing?' Pug indicated the maps and notes Locklear had brought from the barn at Yellow Mule. 'There are at least a half-dozen false messages here, to be conveniently found by the Prince's agents, all stressing attacks at unlikely places, Tannerus, Eggly, Highcastle, even Romney.'

Gorath's head came up. 'I have heard that name.'

'Romney?' said Arutha. 'What do you mean?'

'Only that I have heard Romney mentioned by those who are in service to Delekhan. There are agents working for him in that area.'

'Would you know them?'

Gorath shook his head in the negative. 'Only a few close to Delekhan might know who is working for him among the humans: Nago's brother Narab, his seniormost advisor, his son Moraeulf, and the Six.'

'Who are the Six?' asked Pug. 'You've mentioned them before.'

'No one knows. They are swathed in robes as dark as those of your Tsurani friend and yourself, with deep hoods.'

'Pantathians?' suggested Pug.

'Not snake priests, I'm certain,' said Gorath. 'They speak as you or I do, though there is an accent to their voices. Yet, they serve Delekhan and give him the might to unite the clans. Their magic was powerful enough to force Nago and Narab to heel on two occasions when they sought to distance themselves from Delekhan. And among our people, they were the mightiest of Spellweavers.'

Arutha said, 'Pug, would you bring that map over here, please?'

Pug got the map indicated by the Prince, one of the central third of the Kingdom. He placed it on the table next to the one brought by Locklear from Yellow Mule. 'What cause would Delekhan have to operate out of a river town in the heart of the Kingdom?'

Pug said, 'Perhaps *because* it is in the heart of the Kingdom?' He pointed to the location. 'When Murmandamus came against us, he moved through Highcastle, and crossed the High Wold, moving to the southwest to enter the Dimwood and strike south to Sethanon. What if this time Delekhan ran this pass here, and came down the River Cheston by barge?'

Arutha nodded. 'At Romney he could turn to the River Silden and north of the City of Silden he could turn westward and force march to Sethanon. It's his fastest route and his easiest if I've got the Armies of the West tied up at LaMut and Tannerus and a dozen other places from here to Yabon. He'd be west of the King's Armies, too.'

Arutha looked at Gorath. 'At last something starts to make sense.'

Gorath said, 'If I go to Romney, I may be able to find you the proof.'

Arutha said, 'It's a long step from belief to trust, Gorath. Our people have been enemies too many years for trust to come easily.'

'Send me with your soldiers, then,' said Gorath. 'Delekhan must be stopped. If you blunt his attack, send him back to the north with his nose bloodied, his own supporters will throw him down and my people will be saved. As will yours.'

Arutha considered, and said, 'I've got just the person to put on this task. But Jimmy is out conducting some other business for me right now – '

'Nighthawks?' asked Gorath.

'What do you know of that?' asked Arutha.

Gorath explained the encounters in the sewers with the false Nighthawks and Squire James.

Arutha nodded. 'Someone's anxious for me to send the army into the sewers, cleaning out the Mockers while I'm at it. The two things may be related, but they also may be coincidence.'

'I think them related,' said Gorath. 'I have not heard anyone speak of the Nighthawks, but I have heard them speak of Delekhan having sources of information throughout the Kingdom.'

Makala said, 'And from what young Locklear said, agents working within the Empire, as well.' He held up the ruby Locklear had returned to him. 'These thefts have been under way for some time now.' The Tsurani magician looked at Arutha. 'I think these events are all somehow related.'

Arutha nodded. He looked at Gorath and said, 'I'm returning you to your room, under guard. I'll send for you in the morning and we'll map out your journey to Romney. It's weeks out and back by even the fastest horses, and we need information in as timely a fashion as possible.'

Gorath rose from the table and with a slight nod to Gamina and Pug, he left the room.

Arutha let out a sigh of frustration. 'So much of what he knows is based upon overheard conversations and rumour. I believe his warning is sincere, but is it accurate?'

Knight-Marshal Gardan, who had remained silent while the moredhel was in the room, said, 'I don't trust him, Highness. We fought them too often over the years to trust any of them.'

'But what other choice do I have, Marshal?' asked Arutha. 'If his warning is true, we face another Great Uprising, and if we don't guess right, we may be in the same situation as we were last time, with armies racing to converge on Sethanon, with the moredhel already there.'

'Why Sethanon?' asked Makala, looking at the map. 'Why do they believe this Murmandamus is imprisoned there?'

Arutha glanced at Pug and said, 'That is where he disappeared. There's some rumour about the place, and Murmandamus was deluded into thinking that if he seized that city he would cut the Kingdom in half and defeat us.'

It was a weak lie, and Pug knew it, but Makala said, 'Often in war decisions are based on bad assumptions. Is there some proof, though, that Murmandamus is indeed dead?'

'Only my word,' said Arutha. 'For I was the one who killed him.'

Makala looked at Arutha and said, 'And we can pretty safely assume they will not take you at your word, correct?'

Arutha nodded.

Pug shook his head in frustration. 'My daughter and I must leave for a while, but we'll be back, Arutha. I am more concerned by these mysterious six magicians than all the other news brought by Gorath.'

'Yes,' said Makala. 'The mysterious magic-users. We of the Assembly will be glad to aid you, Pug, if we can. Just call on us.'

Pug asked, 'Are you coming to Stardock?'

Makala said, 'I have some messages to pen to those back on Kelewan. I will join you at Stardock soon.'

Pug nodded, took an orb out of his robe and placed his arm around his daughter's waist. He activated the orb, and, with a buzz, the pair vanished.

Arutha said, 'Would that the rest of us could flit from place to place with such speed.'

Makala said, 'To prevent armies from doing just that is one reason my brotherhood of magicians guards these devices so, Highness. We shall have to be cautious of our part in such matters, though given our Emperor's disposition – ' he referred to the fact that Ichindar, Emperor of Tsuranuanni, was in favour of close ties with the Kingdom ' – you can expect aid quickly, should you need it.'

Arutha gave him thanks, and Makala and Gardan departed. Prince Arutha sat in his chamber late into the night, weighing and judging the warning brought by the renegade moredhel chieftain, and no matter how often he wanted to put it aside as a charade, or nothing more than the jockeying for domination between factions in the moredhel nations, he couldn't. Another war was coming; he could feel it in his bones.

Unless his prized agent, a former thief turned courtier, could somehow steal that war out of the very hands of those who sought to unleash it.

Arutha picked up a bell on his table and rang it. Instantly a page appeared at the door. 'Highness?'

'Send word to the guard to notify me the moment Seigneur James returns to the palace, whatever the hour.'

'Highness,' said the page, closing the door as he left to do as he was bid.

Arutha still didn't return to his own quarters, for even having made the decision to send Jimmy with Gorath to Romney, he had a thousand other questions to consider, and foremost among them was 'Who are the Six?'

Gorath was awake as soon as the door began to open. He rose, fists ready, for while unarmed, he was ready to defend himself. He was not confident that no assassin could find his way into the palace. He remembered events from many years ago where the Princess of Krondor almost died at the hands of one employed by Murmandamus.

Gorath relaxed when he saw his visitor was Squire James. 'Greetings,' said the young man.

'Greetings,' said Gorath. He sat down in a chair next to a window overlooking a garden. 'Am I to be questioned again?'

'No,' said James. 'We're going on a trip to Romney.'

Gorath rose. 'As I have nothing to pack, I am ready.'

'Provisions are prepared, though we will be travelling light.'

Gorath said, 'I expected an escort of at least a full company against attack on the road to Romney.'

James smiled and said, 'Too much noise and bother.' He reached into his tunic and pulled out an odd-looking device, an orb with tiny levers on it that could be adjusted by one's thumb. 'And we're not riding.'

'How are we getting there?' came a voice from behind James.

James turned to find Owyn standing behind him.

'*We* are not going. Gorath and I are. You are staying here or heading home to Timons, as pleases you.'

'I can't stay here,' said Owyn. 'I've got nothing to do and I'm not in the Prince's service. And I can't go back to Timons. What if I'm captured along the way and made to talk?'

James smiled. 'What do you know?'

'I know you're bound for Romney,' said Owyn.

'How do you know that?'

'I know how to read a map and I overheard enough between Gorath and Locklear to know that's where I'd be heading next.'

Owyn hurried on in his pleading, 'Besides, I'm from the east and know my way around back there. I've got cousins in Ran, Cavell and Dolth and have visited Silden and Romney.'

James shook his head as if remembering something, and said, 'Never mind. I seem to recall that Locklear and I made a similar brief to someone who didn't want us along, as well, many years ago. Very well, you can come. It's better to have you underfoot than out of sight and dead, I guess.'

James led them to an empty room in another part of the castle, where weapons and travel items were piled. Gorath picked up one sword and said, 'A lamprey!'

'That's a bloodsucker, all right,' said James, 'but why do you call it that?'

'A name, that's all,' said Gorath. 'My people did not always live in the mountains, human. Once we abided on the shores of the Bitter Sea.' He admired the curve of the blade and weighed the heft of the hilt in his hand. He put the sword back in its scabbard and said, 'I will not ask how you came to possess a blade fashioned by my people.'

James said, 'As you might expect.' He pointed to three backpacks. 'Food and other stores, for we may have to do some travelling, but for the most part I hope we're able to conduct our business quickly and be out of Romney.'

'Where's Locky?' asked Owyn.

'He leaves in an hour on another mission for the Prince. I will meet with him after we're done in Romney. This isn't the only iron in the fire, so to speak, though it may be the most important.'

They picked up their belongings, and Owyn asked, 'Now what?'

Again James produced the orb and said, 'Stand close by. Gorath, place your hand on my shoulder, and Owyn, yours on his.' James put his left hand on Owyn's shoulder, and with the right, activated the orb.

There was a buzz in the air and the room around them seemed to shimmer. Suddenly they were in a different room. 'Where are we?' asked Gorath.

'Malac's Cross.' James crossed to open the door and peered out. 'We are in a building owned by friends of the Prince, and I had best lead, else you may find your head split before you can identify yourself.'

They were on the second floor of a building. As they descended the stairs, a monk in plain grey robes turned a corner and stared openmouthed at them. 'Ah – ' he began.

James held up his hand. 'Tell Abbot Graves we're here, brother.'

The monk turned and hurried off to do as he was bid. James led them into what had obviously once been the common room of an inn. A large man with a short, grey-shot beard hurried over and said, 'Jimmy, you scoundrel! What is all this?' He indicated Gorath and Owyn.

'Hello, Ethan. A person of some consequence desires to see us quickly on our way to the east, and back again. Using that Tsurani device was our fastest start.'

'So you come from Krondor?'

James nodded yes. 'Have you horses we might borrow?'

'No, but I'll send a brother over to Yancy's stable and get three. Care to tell me what this is about?'

'No,' said James. 'Trust me.'

The man named Ethan Graves said, 'We go back a long way, together, lad, to darker days when I was another man. But while I hold your master in high regard, my loyalty now lies exclusively with the temple. If this is some matter of concern to the Temple of Ishap, you should tell me.'

James shrugged. 'If I can, I will, but at this point all I have is conjecture and speculation. Still, let me say that it's time to be wary.'

Graves laughed. 'We are always wary. Why else buy this inn and turn it into an abbey on the fly?'

'Are things . . . well?'

Graves said, 'Go see yourself. You know the spot.'

'Will you have horses ready when we return?'

'And whatever else you need.'

'Just horses. We have our necessaries in hand.' He indicated the packs they carried.

He removed his pack and said to the others, 'Come with me. We'll be back for these in an hour.'

They left the inn and Owyn looked over his shoulder. It was a modest building, two storeys tall, with a stabling yard, a pair of outbuildings near the barn, and a storage shed. It sat on the outskirts of a modest-looking town, which stretched off to the east. Monks of Ishap were hard at work replacing the wooden fence around the end of the property with stone.

'What is all this?' asked Gorath as they walked southward, down a path through some woodlands.

'An abandoned inn, which has been taken over by the Temple of Ishap. They are converting it to an abbey.'

'To what ends?' asked Gorath.

'There's something not too far from here they wish to keep an eye on.'

'Which is?' asked Owyn.

'Something neither of you needs to know about.'

They walked for about ten minutes along a path through the woods. They reached a clearing and Gorath halted, momentarily startled by what he saw. Rising up before them was a statue, perfect in detail, of a recumbent dragon, its head upon the ground, its wings unfolding as if it was just about to rise up.

'What is this?' asked the dark elf. He walked around it, inspecting it closely.

'This is the Oracle of Aal,' said James. He indicated a votive offering plate on the ground before the dragon.

Owyn said, 'I thought it but a legend.'

'Like many legends, one based in truth,' said James. He motioned to the plate. 'Toss in a coin and touch the dragon.'

Owyn fished out a silver coin from his pouch and tossed it into the plate. A moment before it touched the surface of the plate, the coin vanished. Owyn reached out and touched the dragon . . .

And was someplace else. It was a large chamber; immense was more accurate, thought Owyn. Air moved in the chamber with the stately leisure of ages, and before Owyn reared up a dragon of gigantic proportions, the head resting upon the ground larger than the largest waggon Owyn had ever seen. The creature's body was resplendent with gems of all hues. Diamonds predominated, but emeralds, sapphires, rubies and opals formed patterns that swirled on the dragon's back, and made her look as though she wore a shimmering rainbow. It was hard to look away.

'I'm asleep?' asked Owyn.

'In a fashion. But quickly, you tread a dangerous path. What would you ask of the Oracle of Aal?'

'I find myself caught up in something I don't understand,

yet I feel compelled to continue with my companions. Is this wise?'

'At journey's end you shall not be as you are now, nor may you ever return the way you have come. The days ahead of you are filled with hardship and many times to come you will think yourself less significant than you truly are.'

'Can I trust the moredhel, Gorath?'

'He is more than even he knows himself to be. Trust him, though he will not always trust himself. He will become a great champion, even to those who curse his name and will never know of his greatness.'

Suddenly Owyn felt his knees go weak and he faltered. Strong hands grabbed him, holding him upright. He blinked and was again standing before the statue. 'What?'

'Are you all right?' asked Gorath. 'You touched the statue and seemed to falter.'

'I was someplace else,' said Owyn. 'How long was I gone?'

'Gone?' said Gorath. 'You weren't gone. You but touched the statue and staggered slightly, then I grabbed your arm.'

'It seemed longer,' said Owyn.

'It happens that way, sometimes,' said James, touching the stone. He withdrew his hand a moment later. 'Who gets to speak to the Oracle is the Oracle's choice. What did she say to you?'

Owyn glanced at Gorath and James. 'Only that I must trust . . . you both.'

'Did the Oracle say anything useful?' asked James, gripping Owyn by the arm.

'Only that the days ahead are filled with hardship.'

Gorath snorted in contempt. 'As if we need an oracle to tell us that.'

James said, 'Let's get back to the abbey and see if our horses are ready. We still have a fair ride ahead.'

'Where are we bound? Salador?' asked Owyn.

'No, the high road to Silden. It's less travelled and more

dangerous for that, but anyone looking for us is still hanging about outside Krondor, I wager, waiting for us to stick our heads out of the palace. With luck, we will be on the road to Romney before our enemies finally learn we are no longer in the palace.'

Owyn nodded, and as they began their return to the inn-turned-abbey, he cast a backward glance over his shoulder toward the clearing in which the dragon statue rested. There was something he had sensed in his dream state, something he had not spoken of: the Oracle was afraid.

SIX

Journey

The Abbot waved a greeting.

They climbed the trail from the dragon statue to the converted inn, and found Abbot Graves waiting for them. 'You'd better get into town before you leave, James,' he said.

'Why?' asked James, looking for signs of trouble in the Abbot's manner.

'About five minutes after you vanished down that trail, a column of riders came past here, heading into the city.'

James squinted towards town, as if trying to see the riders. 'Something was notable about them, else you wouldn't be remarking on them. What?'

'They wore the King's colours. And unless I don't remember my days as a thief in Krondor, old Guy du Bas-Tyra himself rode at their head.'

'That's something we need to see then,' said James. He motioned for Gorath and Owyn to follow him and started walking toward town. 'We'll be back in a while, Graves.'

The Abbot waved goodbye, and turned back into the building.

They hurried into the town, heading down the main boulevard, and reached the town square. There, a full squad of riders were dismounting and tending to their horses before an inn with a chess piece – a white queen – on its sign. The soldiers were all attired in the livery of the royal house, black trousers and boots, grey tunics over which each wore a scarlet tabard with a white circle and scarlet lion rampant, crowned gold and holding a sword:

the King's coat-of-arms. A line of purple around the edge of the circle and upon the cuffs of the tunic showed these were palace guards, those whose first duty was to the Royal Family. Two guards stood at the door, and one said, 'Easy now, friend. The Duke of Rillanon is taking his ease in the commons and no one goes in until he's left or without the Duke's say-so.'

'Then get yourself inside, soldier, and tell him Seigneur James of Krondor is here on the Prince's business.'

The soldier gave James and his companions an appraising look, then went inside.

A moment later a large man, his grey hair flowing to his shoulders and a black patch over his left eye, appeared before them. He stood with his hand upon the door a moment, then waved them in.

Inside the common room, James and the others could see the soldiers of the King's Royal Guard were efficiently checking out the surroundings.

Guy du Bas-Tyra, Duke of Rillanon and First Counsellor to the King of Isles, waved them to a table where he sat down heavily. 'Get me something to drink!' he shouted, and a soldier detailed to be his orderly hurried to where an intimidated-looking barman waited. The man almost hurt himself on the edge of the bar trying to bring out a tray of pewter jacks. He filled them quickly and ran over to place the first one before Bas-Tyra and then served the others at the table. He said, 'Would m'lord care for something to eat?'

'Later,' said Guy, slowly removing his heavy gloves. 'Something hot for me and my men. Cook up a side of beef.' The innkeeper bowed and backed away, knocking over a chair at the next table, which he quickly righted. Guy looked at James and nodded.

James's brow furrowed but he returned the nod. Duke Guy said, 'So, Arutha is sending you east to snoop around?'

James said, 'That's one way of putting it, Your Grace.'

Guy pointed to Gorath. 'Now, explain to me why I shouldn't cut his heart out and hang you for being a black-hearted rogue and traitor to the Crown?'

Gorath's hand tightened slightly on the hilt of his sword, but he didn't move. Owyn's face drained of colour, but he saw James smile.

'Because it would irritate Arutha?'

Guy laughed. 'You haven't lost any of that mouth, have you, Jimmy?'

The young man said, 'I probably never will. We've been through too much for you to seriously wonder about where my loyalties lie, so I judged you were taking out your bad temper on me because you couldn't take it out on Arutha. Why's he got you so peeved?'

The Duke of Rillanon, most powerful noble in the Kingdom after the Royal Family, leaned back in his plain wooden chair and made an all-encompassing gesture around him. 'This. Because I'm here in a town whose only excuse for existing is its location between Krondor and Salador, and because Lyam is concerned about reports that have been coming to the court of renegade moredhel – ' he locked his one good eye on Gorath ' – and some other unsavoury types running loose between here and Romney.'

'Why you?'

'A variety of reasons,' said the Duke. He took a long drink of ale. 'I usually don't drink this early in the day, but I'm usually not riding all night, either.'

'Those other unsavoury types wouldn't be Nighthawks, would they?' asked James.

'They might be,' said Guy. 'What's Arutha hearing?'

'Nothing until I get back and report,' said James. 'But on the way into the city, Locky and these two found a pair of frauds playing the part of the Guild of Death.'

Guy looked off into the distance for a moment, as if weighing his words. 'If you were trying to revive the Nighthawks,' he posed to James, 'and you wanted someone to think that you weren't, how useful would it be to have a bunch of bunglers found out as false Nighthawks?'

James's eyes widened. 'Brilliant! It would take attention off what I was really trying to do, I would have some pawns to offer up as a sacrifice, and the people I'm the most worried about wouldn't take me seriously.'

'Look deep, Jimmy,' said Guy. 'Find who's really behind the troubles we're having. There's an old axiom: *absent any contrary information, assume your enemy will act intelligently.* The corollary of that is: *act stupid, and your enemy won't take you seriously.*'

James said, 'You still haven't said why you're here.'

Guy nodded. 'The King wanted me to personally take a hand in this region. It seems some of our local nobles are suspect.'

'Of treason?'

'Not that, though it might be a remote possibility.' Guy finished his ale. 'Rather, they're suspected of incompetence. My lord, the Earl of Romney, has a guild war about to break out, and seems unable to do anything about it. I ordered a company of Royal Lancers to head that way to bail him out as I left; they should be arriving some time in the next week.'

'What sort of war?' asked James.

'I don't know the details, but it seems the Brotherhood of Riverpullers has raised prices to the point where merchants can't afford to have their goods hauled up or down the river, and the other guilds are lining up in opposition to the Riverpullers. Both sides are hiring swords, and, for all I know, the Earl of Romney has declared martial law by now. Hell, for all I know, the city is in flaming ruin.' He slapped the table for emphasis, as if he didn't care whether Romney did go up in flames.

'And besides this tour to get things in hand, we're showing the King is personally interested; the banners are waving in front of people who need to see them, and I'm also required by His Majesty to give a lecture tonight.'

'Lecture?' asked James, barely able to keep from laughing. 'On what? And to whom?'

Guy sighed. 'On the Battle of Armengar, and to anyone who wants to listen.' He shook his head as if he couldn't believe his own words. 'You know that thug, Graves, who the Ishapians sent here to start that new abbey?'

James nodded. 'I knew Ethan before he heard the call. He was a rough one in those days; one of the better bashers in the Mockers.'

'I can believe it. In any event, he has decided, or rather the Ishapian Temple in Rillanon has decided, that a school is to be built here, in Malac's Cross, "the centre of the Kingdom", and that young nobles are to attend. They call it a *collegium*.' He lowered his voice a bit. 'I think they're distrustful of what our friend Pug is doing down in Stardock and think they may gain by having a similar venue for influencing the young nobles of the Kingdom. And I think they also want a base near . . .' His eyes flickered from Gorath to Owyn, and he let his words lapse.

James knew what he was about to say: near Sethanon and the Lifestone, so all he did was nod slightly. Glancing around the room, James said, 'I don't notice a lot of young nobles in the area, Guy.'

Guy reached across the table and attempted to give James a playful slap to the head, which James adroitly avoided. 'You always were a smart mouth, Jimmy, and you always will be a smart mouth.' James grinned. 'Even if you some day get your wish and connive to be named Duke of Krondor, you'll still be a smart mouth.'

James laughed. 'Maybe. Now, where are these young nobles?'

Guy sighed. 'A few will wander in from surrounding estates, no doubt. That's why I rode all night to get here. Damn weather had my ship reaching Salador two days late, so I'm riding through the night so as not to make a liar out of the King.' He took another sip of ale. 'And it's why I want you to attend the lecture tonight. It's in a house down near the eastern edge of the town. You won't be able to miss it; it'll be the one with all the Royal Guardsmen standing in front of it.' He stood up, and James did as well, Owyn and Gorath a moment after. 'Lyam asked me to do this while I was making arrangements along the frontier with the Western Realm, and as a favour to the Ishapian Temple. I, loyal duke that I am, could hardly say no to my king. You, loyal squire that you are, cannot say no to me. You will attend the lecture this evening as my claque, seigneur. Now, I'm going to see to my men, and then I plan on getting some sleep.'

The Duke departed, heading upstairs to the rooms set aside for him. Gorath turned to James and said, 'What is a claque?'

James laughed. 'Theatre owners hire them to cheer loudly at performances, to gull the uninformed into thinking the play was brilliant. It gets quite funny sometimes. Five or six people will be cheering wildly, while the rest of the audience is booing and throwing rotten vegetables at the actors.'

Gorath finished his ale and shook his head in dismay. 'Humans.'

The innkeeper came over and said, 'Anything else, gentlemen?'

He studied James's face for a long moment, then said, 'Sorry. Thought you were someone else.' Clearing his throat, he repeated, 'Anything else?'

'If I don't get food on top of that ale, I'll be asleep in an hour,' said Owyn. 'I've never drunk so early in the day.'

Gorath let out a disapproving grunt, but said nothing.

James said, 'Whatever food you're serving, ah . . . ?'

'Ivan's the name, sir,' he said, bowing as he turned to leave.

The door to the tavern opened and three men entered.

James, Gorath and Owyn all looked, their mission making them wary. The three men were locals, and one was carrying a chess set. They set it up on a table and two began to play while the third watched.

Ivan returned and served the food: cold meats, cheese, spiced greens and sweetened apples. He put the platter on the table and said, 'More ale?'

James nodded. Another pair of men came to set up another game of chess and James said, 'This goes on here a lot?'

Ivan nodded. 'The Queen's Row, sir? The chess piece on the sign means something. Old man Bargist, who opened this inn some thirty years ago, was a fair player, and since then, well, travellers and locals alike know that this is where you come to play a match if you want to test your mettle. You play, sir?'

'Not well,' said James. 'My . . . employer plays very well, and has taught me the basics.'

'You can always find someone here willing to play,' he said as he departed to see what the soldiers were ordering.

The door opened and a ragged old woman slipped through, closing it behind her. She came across the room and stood next to James. 'I thought you gone up to Lyton, Lysle. And where did you get those clothes?' She felt the fabric of his tunic at the shoulder. 'Must have snitched them off a baron, from the feel of them.' She squinted at James as if she had trouble seeing him.

'I think you have mistaken me for someone else. My name is James – '

'James, is it?' she interrupted 'Well, then, if it's James, then it's James.' She nudged him with her elbow and winked. 'Like the time you chewed soap and walked around foaming at the mouth, eh, dearie? Taking alms from the gullible? If you say so. Be a love and buy old Petrumh some food, will you?' She then noticed Gorath and said, 'What are you doing with an elf, boy? Don't you know they're bad luck? They're the ones killed my old man Jack, and they're causing all that trouble up at Sethanon. What are you thinking?'

James asked, 'What trouble in Sethanon?'

She leaned down, and blinked, studying James's face. 'You're not Lysle!' she said. Slapping him weakly on the shoulder she said, 'What are you doing with his face?' Her hands came up to her mouth and she stepped back. 'Ow!' she cried. 'You're an evil fairy, that's it! You've taken Lysle's form to trick me!'

James put up his hands. 'Madam! We are not evil fairies.'

'And I am not an elf!' grumbled Gorath under his breath.

The old woman leaned forward again. 'Well, you don't look evil, that's a fact. But you could be Lysle's twin, and that's also a fact.'

James waved Ivan over and gave him a coin. 'See the woman gets some food,' he instructed. To Petrumh, he said, 'You say this Lysle has gone to Lyton?'

'Left a few days ago, he did,' she agreed. 'Said he was to meet a gentleman there. I suspect he'll be in trouble, sorry to say. That's Lysle. Has a knack for trouble, he does. And I doubt the bloke he's meeting is a gentleman.'

Ivan took the woman by the elbow and moved her to a corner table and sat her down before food. She dug in without looking up and James turned his attention to his companions.

'A double?' said Gorath.

Owyn said, 'Could someone have put a lookalike on the road to Romney ahead of us?'

James shrugged. 'Maybe. It's been done before. I saw a double of the Prince years ago in the sewers of Krondor. If it hadn't have been for muddy boots, he might have convinced people he was Arutha and played havoc with things for a while.' He shook his head and said, 'But I doubt it. From what the old woman said, this fellow Lysle's been around here a while. It may just be a coincidence. A while back some fellows up in Tannerus kept trying to beat me up for something done by some other fellow until I convinced them I wasn't that person. Twice, in less than a year, makes me think there's someone walking around who resembles me, and from what I've heard, he's not doing me any favours with the resemblance.' He waved the innkeeper over.

'Have you seen me before?' asked James.

'Can't say as I have,' said Ivan.

'But you thought you had, earlier,' observed James.

'No, I said I thought you were someone else.'

'Who?' asked James.

'Lysle Riggers,' said Ivan. 'Local scoundrel, truth to tell. Has his hands in a lot of . . . questionable activities. Still, he's also a good man to know if you need something done, if you know what I mean.'

'I do,' said James. 'Have you known this fellow long?'

'On and off, off and on, maybe ten years or so,' said Ivan. 'He comes and he goes. Sometimes he's here for a month, then gone a year, other times he's here a year, then gone a month. Never can say what he's up to.' He looked at all three in turn. 'Can I get you anything else?'

'No, that will be all,' said James.

'What now?' asked Owyn, yawning from the effects of drinking early in the day.

James said, 'I'm heading back to chat with my old friend Graves. You might do well with a nap. And tonight we'll go listen to Duke Guy lecture local youths about the Battle of Armengar.'

Gorath said, 'I may stay here. I already know about Armengar. I was there.'

James grinned. 'So was I. But we'll go. It's not politic to disappoint a duke, my friend. It can cause troubles if you do.'

Gorath's answer was an inarticulate snort, but he stood

and said, 'I am going to scout around. From what the old woman said, some of my people may have been nearby. I will look for any signs.'

'Good,' said James, standing up. 'We all have something to do.'

James and Gorath left, and Owyn went to where Ivan stood cleaning glasses behind the bar. 'Could I rent a room for tonight?' he asked.

'Normally I'd be happy to oblige,' said Ivan. 'But the King's men have taken them all.'

Owyn asked, 'Is there another inn nearby?'

'There's one a half day's ride to the west, though I wouldn't recommend it. And there's another a half day's ride to the east, but I wouldn't recommend it, either.'

'Perhaps a spot in your barn?'

'King's men won't allow it, lad. Sorry.'

Owyn turned away and decided to catch up with James. If he couldn't nap, perhaps he might find something interesting to study at the Ishapian abbey.

Much to James's surprise, there were a fair number of attendees at Guy's lecture on the Battle of Armengar. Owyn sat nearby looking sleepy. He had returned to the makeshift abbey and managed to find some books to read. He had become engrossed in one that touched on magic and found several things of interest.

During the talk James had elbowed Owyn twice as the young man was about to drop off to sleep. As the lecture wrapped up, James was forced to admit the old commander of the defences at Armengar had been truly brilliant in his tactics. The mere fact that a fair number of survivors reached Yabon safely while the Teeth of the World had swarmed with goblins and moredhel looking for human refugees had been an accomplishment.

The audience applauded politely when Guy concluded his

remarks and several young nobles from the area approached to talk to the Duke. James said, 'Wait here,' and went to make his goodbyes to Guy.

When he returned, he said, 'Let's go.'

'Where?' asked Owyn. 'There's nowhere to sleep in town.'

'We can sleep on the floor at the abbey and get a fresh start in the morning.'

'Good,' said Owyn yawning. 'I'm all in.'

'You'd better learn to hold your ale, Owyn,' said Jimmy with a grin.

They moved down the road and James wasn't surprised to find Gorath suddenly beside them, though Owyn nearly leaped sideways when the dark elf materialized out of the evening gloom.

'Find anything?' asked James.

'Tracks. Moredhel have been through here recently.'

'What else?'

'A fair number of people are passing to the north of town, not through it.'

'We can assume they don't wish to be seen. Which way were they going, east or west?'

'Both. A lot of people moving in both directions, but keeping out of sight.'

James shook his head in consternation. 'Damn, I don't like any of this.'

They remained silent as they reached the abbey. 'Well,' said Graves, as they entered the former common room, 'how was the lecture?'

'Could have used a singer,' said James with a straight face.

'Duke Armand de Sevigny will be lecturing here next month,' said Graves, 'and Baldwin de la Troville the month after.'

James assured him, 'I'll try not to miss the lectures. Have you a place we can sleep tonight?'

'You're welcome to bed down under the tables here in the commons, Jimmy; but the rooms upstairs are being used by the brothers or for storage.'

'Under the tables will be fine,' said Owyn, unrolling his bedding from his travel bundle. Gorath did likewise without comment.

James sat opposite the thief-turned-cleric and, keeping his voice low, said, 'Why here, Ethan?'

The Abbot shrugged. 'I don't know, Jimmy. You know the order wants to be close to Sethanon,' he said. 'There's a rough village forming up a few miles south of the old city but nothing you'd call a proper town. It's still a decent trading route, though, and some folks seek to profit by caravans and traders passing by. It would be too obvious for us to try to build an abbey there. But here we can be circumspect and still send a brother up there to snoop around from time to time, just to ensure nothing disturbs the status quo.'

'I noticed the next two lecturers are men Bas-Tyra trusts.'

Graves nodded. 'There're too many strange things going on for him to do otherwise. Some of the other nobles . . .' he shrugged. 'They're not as trustworthy as they could be.'

'You don't think treason, do you?'

'I don't know what to think,' said Graves. 'I'm a former thief who has been handpicked by the temple in Rillanon for a potentially difficult, even critical assignment.' He looked down as if afraid to look James in the eyes. 'I don't know if I'm equal to the task.'

'I've never heard you act the shy role before, Ethan.'

Graves sighed. 'There's a lot you don't know, Jimmy. I have some old . . . ties, you could say. They aren't easily broken. You know how it is.'

James laughed. 'Better than most. I have a death mark on me from the Mockers if I cross their boundaries, yet I do so all the time. And they conveniently ignore my trespass when it suits them. I know what you mean, I think.'

Graves said, 'I hope when the time comes you do know what I mean.' He stood up. 'I must retire. There's a great deal to be done around here. Have a good night, Jimmy.'

'You too, Ethan.'

James undid his own bedding and lay down next to Owyn, who was already fast asleep. As he drifted off himself, he wondered just what Graves had meant by 'when the time comes'.

The north wind cut through the night. James huddled under his cloak as the three stayed close to their fire. The road from Malac's Cross to Silden was far less travelled than the King's Highway to Salador, but it was a more direct route. Behind them, the three horses James had purchased, along with tack, were quietly eating the grain he had bought for them.

Owyn said, 'James, I'm worried about something and I've been meaning to talk to you about it since we left Malac's Cross.'

Gorath said, 'You have seemed troubled.'

James asked, 'What is it?'

'I don't know exactly, but it's something I picked up from the Oracle . . . A sense of foreboding.'

'Given our circumstances,' said Gorath, 'that is not particularly inappropriate.'

'What do you mean?' asked James, looking intently at Owyn.

'It's like the Oracle was worried.'

James was silent, then said, 'I'm no expert, and I've never seen the Oracle myself, but from what I hear, the Oracle can tell futures, but not her own.'

'Futures?' said Gorath.

James paused, then said, 'Maybe I'm telling it wrong, but the magician Pug once told me that the future is not set in stone, but the result of many acts, and that by changing an act today, the future changes.'

Owyn said, 'As if you had not come south, Gorath, Delekhan's plans would be further along.'

'I understand that,' said Gorath. 'But if the future is fluid, what good is an Oracle?'

James shrugged. 'There is a great deal of wisdom in this Oracle, I have been told.'

Owyn looked at Gorath and said, 'I think James is right. But I still don't know about that feeling of worry.'

'Perhaps the Oracle's fate is bound up in what we do,' suggested Gorath. 'Then it would be difficult for her to see the future, if what James said is correct. Perhaps that is the source of the worry.'

James said nothing. He was one of the few who knew of the existence of the Lifestone under Sethanon. Only a handful of those who had been at the battle knew of the magic relic from the time of the Dragon Lords. Few knew that the Oracle of Aal was the guardian of the Lifestone and resided in a vast chamber below the City of Sethanon.

The statue at Malac's Cross was designed to mislead those who knew nothing of the dragon Oracle's existence. Should any come seeking the Oracle, it provided the means for them to contact her without actually being in her presence.

James said, 'I'm trying to puzzle together some things. We have Tsurani Great Ones getting their riches stolen, so that Tsurani renegades can sell them to moredhel raiders, who swap them for weapons. We have a false Guild of Death, maybe to mask some real Nighthawks who survived the night we burned their headquarters to the ground in Krondor, and a lot of false trails in the west predicting an invasion from the north.'

Gorath said, 'My people will proceed cautiously. They will want some indication from Delekhan that Murmandamus indeed is alive in Sethanon, being held there against his will, before they will march.'

James said, 'No offence to your people, but that sort of "proof" is easy to make.'

'Agreed,' said Gorath, 'which is why Delekhan is attempting to remove all of us who were in opposition to him.'

James lay back, wrapping the cloak around him. 'Well, we may find answers or we may not, but right now I could use some sleep.'

'You going to look for that double of yours in Lyton?' asked Owyn.

'It's on the way,' said James. 'Might as well while we're passing through town.'

Owyn rolled over, trying to get close enough to the fire to stay warm without burning himself. Gorath just lay silently, until he was asleep.

Sleep was a long time in coming for James as he wrestled with all the fragments and clues he had. Somewhere in all this apparent chaos was a pattern; somehow all the pieces came together and made sense.

The ride to Lyton was uneventful until they reached the outskirts of the town at sundown. Off the side of the road stood a forlorn farm, abandoned by the look of it, with a ramshackle barn, around which skulked black-clad figures.

Gorath saw them first, and James said, 'I wouldn't have even noticed them if you hadn't pointed them out to me.'

'There are four of them, and they seem very curious as to the contents of that abandoned barn,' said the dark elf.

James said, 'My bump of trouble is itching like mad. I think we've found our real Nighthawks.'

Owyn said, 'What do we do?'

Pulling out his sword, James replied, 'Kill them before they notice us, if we're lucky.'

He turned his horse off the road and moved forward at a trot. They travelled across an abandoned field overgrown with tall grass which rose to chest height on the horses. It masked their movement for a while, as the dark-clad figures seemed intent on the barn, which allowed James and his companions to reach the edge of the field before being seen.

The assassin who first saw them shouted and two others turned, as James spurred his horse forward to charge. One of the black-clad men carried a sword and readied himself to strike at James, while another leaped out of the way. At the corner of the barn, a third figure easily drew an arrow and nocked it to his bow, pulling back in a fluid draw. Suddenly a dark nimbus of energy splashed the side of the barn, missing him, but distracting him enough that he fell back without shooting.

Gorath upon the second man was leaping from the back of his horse, while James engaged the first. Owyn cursed as he realized that while he had managed to unravel the mystery of the spell Nago had thrown at him and could now duplicate it, he couldn't control it very well. He hoisted his staff over his head like a war club and rode toward the bowman, trying to strike him before he could loose his arrow.

Gorath crushed his opponent's throat with the flat of his blade, and rose up to see James having difficulty with his man, while Owyn rode around flailing at the third with his staff. The bowman was so busy trying to keep from having his head stove in by Owyn's staff he couldn't stop long enough to shoot. He finally tossed down his bow and tried to draw his sword.

James saw Gorath standing uncertain of which way to move, and shouted, 'Find the fourth one!'

Gorath was off without another word, moving around the corner to find the door of the barn open. Inside was darkness to confound the human eye, but to the dark elf it was a pattern of darkness and light, greys and darker greys. He saw movement in the rafters above and along one wall to his left. He waited.

A moment later the figure in the rafters slipped, causing some hay to fall, and the figure near the wall let fly with an arrow in the direction of the sound. Gorath charged. Before the Nighthawk could pull and fire again, the dark elf was upon him.

The struggle lasted mere seconds as Gorath quickly killed his man. Outside Jimmy bested his own and turned his attention to the one Owyn harried.

When the fighting was over, James and Owyn entered the barn and James said, 'What's here?'

Gorath pointed up to the rafters and announced, 'Someone hides up there.'

James said, 'Come down. We mean you no harm.'

A man lowered himself from the rafters, hanging by his hands a moment before releasing his hold and dropping to the dirt floor. He landed nimbly on his toes and looked at his rescuers. 'Thanks,' he said.

The man moved toward them and when he stopped a few feet away, Owyn said, 'Gods!'

James looked at the man, who looked enough like him to be his twin. 'You must be Lysle,' said James.

'Why do you assume that?' asked the man.

'Because people keep mistaking me for you,' said James, moving around so he faced the door and the scant light from outside could strike his features. 'It got me almost murdered by some unhappy folks up in Tannerus some months back.'

The man laughed. 'Sorry, but they're waiting for me to return with some items they sent me to purchase on their behalf. I've been distracted and am overdue in getting back there.' He paused a moment, then said, 'You do look enough like me to confuse people, it's true. I'm Lysle Riggers.'

'I'm James, from Krondor,' came the reply. 'These are my friends, Owyn and Gorath. We were on our way to Romney and when we were in Malac's Cross an old woman thought I was you.'

'Old Petrumh,' said Lysle. 'She's a little crazy. She's been that way since her husband died in a fire. Most of the folks in town give her something to eat or let her sleep in their barns. For some reason she's taken to telling everyone she's my gran.' He shook his head.

'Care to tell us why a bunch of Nighthawks are trying to kill you?'

'Nighthawks?' asked Riggers, shrugging. 'Assassins? Can't say as I would know why. Maybe they thought I was you.'

Gorath looked at James and said nothing. Owyn said, 'Maybe – '

James cut him off. 'No, someone wants *you* dead, Riggers. Let's head into town and maybe by the time we get there you'll remember why.'

The man looked at the three before him as if weighing the possibility of flight or resistance. Obviously discarding either as an option, he nodded. 'Let's go. The Wayside is a decent enough inn, and I could use an ale after all this.'

'Check the bodies,' said James. Gorath and Owyn went outside to do so. 'You have anything you need to fetch?' asked James.

Riggers said, 'No. I had a sword, but lost it somewhere back in the woods running from those four. It wasn't a very good one. I'll take one off the dead outside.'

Moving outside, James said, 'Fair enough.'

Owyn said, 'They're carrying nothing, James. No papers, no money, nothing. Just weapons and those black clothes.'

Gorath came over and said, 'And these,' as he held up a medallion with a hawk on it.

James took it, inspected it, and threw it to the ground. 'These are the real Nighthawks,' he said. 'Not those frauds down in Krondor.'

'Frauds?' asked Riggers.

'It's a long story.'

'Good,' said Riggers. 'That means a second ale. Let's go.' He set off toward the distant lights of the town, while the others mounted up.

Owyn rode next to Jimmy and said, 'For a fellow who

was about to be chopped up by assassins, he's pretty cheerful.'

'Yes, he is,' said James.

They followed their new acquaintance into town.

SEVEN

Murders

The inn was crowded.

Lysle Riggers led James, Gorath and Owyn into The Wayside, a tavern whose location was reflected in its name, situated as it was just on the edge of the city, and a good walk from the main street. But it seemed a popular place, with workingmen, armed fighters and some unsavoury-looking sorts packing the common room.

James and his companions had left their horses with the lad who worked in the stabling yard, giving instructions for their care, and followed Lysle inside.

Lysle led them over to a table in the corner. He motioned for them to sit and waved to the barman, who hurried over to take their order. James ordered a round of ale and some food, and the barman offered a quick glance between him and Riggers, but said nothing as he headed back to the kitchen.

Riggers said, 'Well, then, I owe you a story.'

'A long one, you said,' observed Gorath.

Riggers said, 'And you shall have it, but I have one question. What brought you so fortuitously to my rescue?' He studied James a moment, then said, 'If it was pure chance, then fate has a curious sense of humour, my friend.'

James said, 'It was chance of a sort, though I had heard your name down in Malac's Cross, as a few people seemed to think I was you. As to how we came to your rescue, that *was* pure chance, though we were on the lookout for just the sort of trouble you found yourself in.'

'You recognized my assailants,' Riggers said, lowering his voice. 'Obviously you know more than the average mercenary.' He jerked his head toward Gorath. 'His kind have been seen around here in increasing numbers lately, though rarely openly walking around with humans. All of which leads me to think you're someone about whom I need to know more before I launch into my long story.'

James grinned. Riggers returned the grin and again the others were struck at the resemblance. Owyn said, 'If you're not brothers, the gods have a fine sense of whimsy.'

'That they do,' said Riggers, 'irrespective of any other thing.'

James said, 'Here's what I can tell you. I'm working for people who presently have no reason to want you dead, Riggers. Let's not give them one. They are also people who are at odds with those employing your would-be killers.'

'*And the enemy of my enemy is my friend*,' said Riggers, quoting the old truism.

'To a point,' said James. 'At this time I like to think we may have more reasons to help one another than not.'

Riggers was silent for a minute, then the food arrived, giving him another moment of respite as he took a slice of cheese and laid it over warm bread. After the ale arrived and he took a long pull on his mug, he said, 'Allow me to be a little circumspect, and I'll tell you what I can.

'I represent interests in Krondor, well established and well connected. They have trading relationships throughout the Kingdom, and into Kesh and across the Bitter Sea to Natal. Lately they've been harried by a new competitor, who seeks to disrupt established business relationships and carve out a new trading empire.'

James considered this a moment, then said, 'Care to name your principals or your new competition?'

Lysle's grin stayed in place, but the humour left his eyes. 'No, to the first, but the second is a personage of some mystery. He's called "the Crawler" by some.'

James leaned forward and spoke low enough that only those at the table might hear him. 'I'm Seigneur James, of the Prince's court, so I'm the King's man. But I was also known for a time as Jimmy the Hand, so I know of whom you speak. "There's a Party at Mother's".'

'"And a good time will be had by all,"' finished Riggers. 'You're Jimmy the Hand? I never would have believed it.' He sat back. 'I don't visit Krondor much. My . . . employer prefers I stay out here in the east. But tales of your rise have travelled far and wide.'

'It may be we have more in common than you know,' said James. He told of the false Nighthawks in the sewers of Krondor and the suspicion that someone was trying to finesse the Prince into raiding the Mockers' hideouts in an attempt to find those false Nighthawks.

'That sounds like the Crawler,' said Lysle. 'He would happily set Crown against Mockers, and sit back and watch. If the Mockers somehow survived, they would be weakened enough that they couldn't oppose him; if they were destroyed, he could move in and take their place.'

'That's unlikely as long as Arutha's in Krondor,' said James. 'He's too savvy to get sucked into that obvious a ploy. What *is* of real concern to us is the existence of these genuine Nighthawks, the ones who were seeking to separate your head from your shoulders.'

'I won't even ask why,' said Lysle. 'I'll assume that it has something to do with the good of the Kingdom.'

'They had a strong hand in repeated attempts to kill Prince Arutha ten years ago. If they're the survivors of that first bunch, or someone else is attempting to trade on their reputation, either way they're a menace. What can you tell us about them?'

Lysle sat back. 'I'm off for Tannerus in the morning – to put right that little matter that almost got you beaten to death when you were last there – so I'll tell you what I know. There's two places this Crawler seems to have taken a foothold. I hear he's got a lot of the crime on the docks in Durbin under his control, and he's dislodged the locals over in Silden. The Mockers were never strong outside Krondor, but they always had good working relationships throughout the Bitter Sea, and a lot of influence in Silden. Lately problems in several Bitter Sea ports have put a crimp on Mocker business and those friendly to the Mockers have vanished in Silden. But the real pot about to boil over is up north; there's a lot of confusion in Romney right now, and from what I can gather, a lot of this Nighthawk business is being run through there.'

'We've heard of some problems there.'

'The Riverpullers' Guild?' asked Lysle.

James nodded.

'That's the Crawler,' continued Lysle. 'He starts at the docks, making it difficult for cargo to get in and out of a city, and wears down both the merchants and local thieves. After a while, people start paying protection to get their goods in, and once he's in their pockets, he never leaves. Damon Reeves is the head of the Riverpullers, and he's an honest man, but someone near him has been whispering in his ear.'

James said, 'You think this Crawler is behind the revived Nighthawks?'

'I don't know what to think. He may have tired of me flitting around causing him troubles and put a price on my head. Or he might be behind them. Or it might be someone else wants me dead for entirely different reasons. I've made a few enemies in my time.' Lysle grinned at that.

'I have no doubt,' said Gorath, dryly.

'Where should we start?' asked James.

'Start with a man named Michael Waylander. He's always at the centre of these problems, it seems. Arle Steelsoul, of the Ironmongers, is leading the opposition to the Riverpullers. Both sides, at least, will talk to Waylander. It's rumoured he has his hands in a couple of shady things; nothing too important, but enough to make him dangerous.'

'Anything besides that?'

'Nothing I care to share with you, but also nothing that kept from you will hinder your efforts.'

'Well,' said James, 'it's more than we had before we ran across you. If you're off for Tannerus tomorrow, we'll know where to find you.'

Lysle grinned and James felt as if he was looking in a mirror. While Lysle was two or three years older than James, the likeness was uncanny. 'That's where I'm heading now. Who knows where I'll be if you come there looking for me?'

James fixed him with a knowing gaze and said, 'Trust me, my friend. Now that I've made your acquaintance, I'll be keeping an eye on you. We'll meet again, have no doubt.'

Lysle finished his food, excused himself and left the three alone. 'I'll see about a room,' said James. He made arrangements and the three retired for the night.

In the morning, they headed for the stabling yard of the inn and discovered a confused stableboy. 'Horses, sir? But last night you took one, and sold my master the other two.'

James turned and looked down the westward road where beyond his vision the village of Tannerus lay. Silently he swore he would certainly find Lysle Riggers again some day. And if any doubt at their being related had existed in James's mind until this minute, it was now completely vanquished. Suddenly laughing, James said, 'Well, I guess we need to buy some horses, lad. What have you to sell us?'

Owyn and Gorath exchanged curious glances at James's strange reaction, but neither said a word as James waited for the boy to fetch the stable-master so he could start haggling to buy three horses.

Armed men had thrown a barrier across the road into Romney, and signalled the three riders to halt. 'What's this?' asked James.

One of the men stepped from behind the barrier, mostly

grain sacks and crates, and said, 'We're not letting strangers into Romney right now.'

James said, 'I'm on the King's business, and I bear warrants from the Prince of Krondor.'

'Prince of Krondor, is it?' said the man, rubbing his chin with his gloved hand. He looked like a stevedore, shirt sleeves rolled up high on his powerful arms, heavy chest and neck, his face burned brown by the sun. He carried a long wrecking bar, the kind used to open heavy crates off-loaded from riverboats, and he looked eager to use it. 'Well, the Prince is a long way away; it's not even the Western Realm, you see, so I can't see as why that cuts any ice with us.'

'Who's in charge here?' said James, jumping down from his horse and handing his reins to Owyn.

'Well, normally it's Michael Waylander, who's trying to keep the Riverpullers from taking over the city, but he's in town right now taking care of some business, so he left me in charge.'

'And your name is . . . ?'

'I'm Karl Widger,' said the man.

Before he could move, James spun on him, hitting him as hard as he could in the stomach. The man went over with a loud 'oof' and James brought his knee upward into Widger's descending face. Karl went down like a dropped brick.

Pointedly stepping over the fallen dockworker, James said, 'Would one of you run into the city and fetch Michael Waylander here? Tell him Karl is incapacitated and there's no one in charge. Unless,' he added, pulling his sword, 'one of you cares to come over here and claim he's now responsible for keeping us out of Romney?'

Two men behind the barricade conferred and one ran off, heading over a small bridge that separated the road into Romney from the King's Highway. None of the others seemed eager to come over the barricade and challenge James, but James knew he couldn't just ride through a dozen armed men.

Owyn dismounted and handed the reins back to James. 'That was bold.'

Under his breath, Jimmy said, 'And a little stupid. I hit that walking tree trunk as hard as I could. Damn near broke my hand, and it was only his stomach. I'm glad I didn't try to hit his head. I'd probably have broken every knuckle. My knee's throbbing like mad.'

It didn't take long for Michael Waylander to arrive. He was a tall man, blond and sporting a short-cropped beard that looked reddish in the afternoon sun. 'What is going on here?' he demanded.

'I might ask you the same thing,' said James. 'I bear warrants from the Prince of Krondor and I'm on the King's business. How dare you bar my way?'

'We're acting under the authority of the Earl of Romney,' said Waylander. 'We've had a lot of trouble lately; damn near a guild war.'

'Guild war?' asked James, as if he had heard nothing about this before.

'Damn Riverpullers are raising prices in violation of every agreement that's in place, and they're threatening to shut down all business up and down the river. I represent an alliance of other guilds: glaziers, rope-makers, carpenters, smiths and most of the local merchants, and we refused to pay.'

James said, 'Let me shorten this for you. You tried to make arrangements to get your own cargo in and out of the city and the Riverpullers started dumping goods in the river and wrecking boats.'

'More,' said Waylander. 'They killed two apprentices three weeks ago and fired a half-dozen boats.'

James said, 'Well, those are local matters. We're on business for the Crown and will brook no more delays.'

'Let me see your warrants,' said Waylander.

James hesitated. This Waylander was no noble or Crown official. By rights he had no legal standing and James was not under any obligation to humour him. But practical considerations and a dozen armed men made him reach into his tunic and pull out his travel warrant and a demand for aid warrant, instructing any noble to aid James in his mission for the Crown.

'Well, we couldn't be too careful. The Riverpullers were hiring swords and the city's become an armed camp. We can't do much about those inside the city already, but we can keep more from coming in.' He handed over the warrants.

'What about the Earl?' asked Owyn. 'Isn't he keeping the peace?'

'We don't have a garrison here, son,' said Waylander, and something in his tone led James to think he liked the idea. 'We're in the heart of the Kingdom and the most trouble we have is the occasional drunken brawl on the docks or a few bandits riding down from the northern hills to ambush someone on the road. We have a city constabulary, but most of those men are on one side or the other in this dispute. The Riverpullers are the most important guild in this area, but the other guilds together are stronger. It's a close thing and we don't have many neutral parties in Romney. Earl Richard asked me to come up from my home in Sloop, a village a half-day's ride south of here, just because I'm not local; I have a lot of friends on both sides of this, and sometimes they'll listen to me. But the Riverpullers are out of line and there's no other way to see it.'

James put his warrants back in his tunic and said, 'I expect they'd have something different to say on that matter. But that's no concern of mine. I need to see the Earl.'

Waylander was about to say something when a clatter of hooves from behind caused James to look. A company of riders was approaching at a leisurely pace up the road, a banner at the head of their column announcing the presence of the Royal Lancers.

Their leader approached, held up his hand for the halt, and said, 'What's this then? Clear the way, you men.'

James nodded, Waylander gave the order, and the men started pulling aside the grain sacks and crates.

James walked to stand before the officer, and after a moment, the officer said, 'What are you looking at, man?'

James grinned. 'Walter of Gyldenholt? So Baldwin sent you south, finally?'

The former captain from the garrison at Highcastle said, 'Do I know you?'

James laughed. 'We met at Highcastle. I'm James, squire of the Prince's court.'

'Ah, yes,' said the old captain. 'Now I recall you.'

James couldn't help but grin. When he had first met the captain, he had been one of the victims of Guy du Bas-Tyra's fall from grace, an officer in service to Guy's most loyal ally, the result of which had been years of hard service with the border barons. Glancing at Walter's girth, he said, 'Peacetime's been good, it seems.'

'What brings you here, squire?' asked the captain, ignoring the friendly barb.

'The Prince has us running some errands for him. You're the company Guy sent here to restore order?'

'We are,' said Walter. 'Would have been here a few days ago, but we ran into a spot of trouble to the south. Band of lads in black objected to our coming this way. Caused us a merry chase, but we managed to kill a few before the rest got away.'

James looked at Owyn and Gorath. 'These are things we had better not speak of in the open, captain. I have to talk to the Earl. I imagine you do as well.'

'Indeed,' said the captain, motioning for his men to move forward, through the barricade now open before them. 'Ride in with us, squire. We'll keep the ruffians off your back.' He smiled at James.

James laughed and mounted his horse, motioning for his companions to join the end of the column. There were fifty lancers in the company, enough to prevent serious trouble, and keep both sides of the dispute from doing anything rash, or at least James hoped so.

Waylander said, 'We were only holding this bridge until the lancers arrived, squire. Tell the Earl my men and I are heading home to Sloop.'

James acknowledged the man's request and they rode across the bridge.

Romney was a major trading centre in the east. The city was big enough to be considered huge by western standards, but here in the eastern half of the Kingdom it was a modest sized place, about half the size of Krondor. With fifty lancers at hand, the Earl could re-form his constables and restore order as long as neither side in the dispute opted for open warfare.

The tension in the city was almost palpable. As they rode in, curious onlookers glanced out of windows or cleared the streets, letting the soldiers pass.

Gorath said, 'There is a lot of fear in the air.'

'People worry when riots break out,' said James. 'Even if you're not taking sides, the violence can sweep you up and carry you into harm's way. Many a man has died trying to explain he wasn't taking sides in a guild riot.'

They rounded a corner and found themselves entering the city's square dominated by a large fountain. James was struck by something odd. 'There aren't any hawkers or vendors about.'

Owyn nodded. 'I've been here before, on my way up to see my uncle in Cavell Keep, and there are always merchants in the main square.'

Gorath said, 'Perhaps they were fearful of being swept up in that violence you spoke of.'

James nodded. A large inn occupied the north side of the square, a black sheep against a green meadow painted on the sign hanging over the door. 'We'll headquarter here,' announced Walter of Gyldenholt.

The lancers dismounted and whatever James might have

thought of the truculent former captain from Highcastle, his squad was the model of efficiency. The captain waved over a passer-by and said, 'Do you know where the Earl of Romney is?'

The man said, 'He's taken up residence in that house there, sir.' He pointed to a house across the square.

Handing the reins of his horse to an orderly, Walter dismounted and said, 'Squire James, let's go call upon his lordship.'

James dismounted and said to Owyn, 'Find us a room, but in a different inn. We'll be able to snoop about a bit easier if we're not keeping company with fifty Royal Lancers.'

Owyn said, 'I know just the place. I stayed here with my father once.' He pointed. 'Down that street is another bridge, crossing the River Cheam, and just on the other side is an inn marked by a green-cat sign. We'll wait for you there.'

James turned and followed Walter, who marched purposefully to the door of the house. He had barely knocked when the door opened and a servant said, 'Enter, sirs.'

The man wore a castle tabard, with the Earl's coat-of-arms on it, a stylized river with a fish jumping from it and over a star. The servant led them to a small parlour at the rear of the house.

Earl Richard was a youthful man, but one who looked more the part of a merchant or tradesman than a noble, despite wearing armour and a sword. James had grown up amidst nobles who were fighting men as well as rulers, and these eastern nobles who wore swords for decoration took some getting used to. The Earl's voice was surprisingly deep and forceful. 'Welcome, gentlemen. My Lord Bas-Tyra answered my request.'

James let Walter speak first. 'We came straight away, sir.'

'How many men did you bring?'

'A full company of fifty Royal Lancers.'

The Earl appeared worried. 'I hope that's enough. I would really prefer to settle this dispute without resorting to force.'

Walter glanced at James and shrugged. The Earl noticed the exchange and said, 'And you are?'

'James, squire to Prince Arutha,' he said, producing his travel warrants and demands for assistance. The second document seemed to produce increased distress in the Earl. 'What sort of assistance?'

'At this point, information, m'lord. We have heard rumours of increased activity in the area by the Brotherhood of the Dark Path, as well as the possibility of a return by the Nighthawks.'

'Possibility?' asked the Earl, his colour rising. 'Doesn't anyone read the reports I forward to the Crown? Of course there's a *possibility*! They've killed two members of the Ironmongers' Guild for the Riverpullers, and killed two members of the Riverpullers, as well; they'll kill for whoever pays them. I hear Baron Cavell is hiding out in Cavell Village because they're stalking him! He lives in a small residence with his household guards in every room.'

Something about Cavell rang familiar in James's memory, but he couldn't put his finger on it.

James said, 'Well, then, m'lord, my companions and I will be around for a few days, asking questions. We'd prefer it if no one else knew our visit was official. If anyone asks, we are here to convey the Prince's greetings while en route to somewhere else.' He glanced at Walter. 'I'll be staying over at the Green Cat Inn, to lend credence to that, captain.'

Walter of Gyldenholt shrugged as if it were of no importance to him. He said, 'My lord, we'll be at your disposal. I'll need to speak with your chief constable in the morning and establish a patrol. As soon as the folks around here see a few of my lads riding around, things will calm down.'

James and the captain excused themselves from the Earl's presence. Outside the door, Walter said, 'Well, squire, we'll have things in hand around here soon enough.'

Again feeling the tension in the air, James said, 'I hope you do, captain. I most sincerely hope you do.'

They parted company and James found his horse, mounted, and rode across the city in the direction Owyn had indicated. As he rode, he studied the city.

Romney was located across all three points of an intersection of three rivers. The River Rom coursed down from the Teeth of the World, near Northwarden, the oldest of the border baronies. At Romney the River Cheam branched off to the southeast, while the Rom continued to run southwesterly, turning southeast again as it neared the coast. James paused at the bridge he faced, which arched over the River Cheam. Something was eating at him, a memory he couldn't quite place, and he knew that it was somehow important. He waited to see if anything bubbled to the surface of his mind, then decided it would come in its own good time.

James moved across the bridge and found this side of the city even more tense than the other. Citizens moved quickly, eyes darting around as if expecting attack from any quarter, and nowhere could any of the usual street hawkers be seen.

He reached the Green Cat Inn and rode around to the back of the stabling yard, where he found Gorath and Owyn waiting for him. 'Why aren't you inside, eating?' asked James as he dismounted.

A terrified-looking stableboy said, 'Sir, my master is unwilling to serve your . . . friend.' He indicated Gorath.

Muttering, 'I wouldn't quite call him a friend,' James tossed his reins to the boy and marched in the rear door of the inn. Owyn and Gorath hesitated a moment, then followed.

Inside, James saw a large man, advancing in years, but still broad of shoulders with imposing muscle under a broad girth, turn to see who entered from the stable yard. He pointed a beefy finger at Gorath and said, 'You! I told you I'll have none of your kind in my inn!'

James hurried to put himself between the innkeeper and Gorath. 'And just what kind would that be?' he asked.

The man looked down at James, appraising him and coming to a halt. The young man was quite a bit smaller, but something in his manner made the barman stop. 'Dark elves! Fifteen years I served on the border, and I've killed enough of his kind to know them. They killed enough of my comrades, as well. And who the hell are you to ask?'

James said, 'I'm Seigneur James, squire to Prince Arutha of Krondor. He's my companion and we're on a mission for the Crown.'

'And I'm the Queen of Banapis,' said the innkeeper in return.

James grinned as he reached into his tunic and produced his warrants. 'Well, Majesty of Love and Beauty, read these, or else I'll have to go fetch Earl Richard to vouch for me, and let's see how much he likes being dragged over here given the temper of the city right now.'

The old man could read, but slowly, with his lips moving. James didn't offer to help him out. After a moment, he handed back the documents. 'Damn, you are some sort of Prince's officer, aren't you?'

James shrugged. 'If I were in the army, I'd be a Knight-Lieutenant, if that makes it easier for an old soldier like you to grasp. Now, I want a room big enough for the three of us, ale, and food.'

The man threw a black look at Gorath and turned his back on James. 'Come this way . . . sir.' He led them to the bar and went behind it. He produced a large iron key and said, 'Top of the stairs, all the way back on the right.' James took the key, when a light entered the man's eyes. 'Six golden sovereigns a night.'

'Six!' said James. 'You thief!'

'It's two per person. Take it or leave it.'

Knowing full well that the fifty lancers would eat up a lot of rooms at local inns, James said, 'We'll take it.'

'In advance.'

James counted out twelve coins and said, 'Two nights. If we stay longer, we'll pay the day after tomorrow.'

The man swept up the coins. 'And that doesn't include the cost of food or ale,' he said.

'I was sure of that,' said James. To Owyn and Gorath he said, 'Let's fetch our kits, then we'll eat.'

They got their travel bags off their horses, ensured the stableboy knew what he was doing, and went upstairs. As James had expected, it was the least desirable room in the inn, at the back over the stable. He decided not to make an issue of it.

Downstairs they endured slow service, even though there wasn't much of a crowd. James was deciding at what point he would have to take the old soldier who ran the place down a peg when the food finally arrived. To James's delight, it was well prepared and of good quality.

As they ate, they discussed the situation. James shared the little information he had with them, and Owyn said, 'So the Nighthawks are working for the Riverpullers or the Ironmongers?'

'Neither,' suggested Gorath. 'Confusion and discord are Delekhan's allies here in the Kingdom.'

'I believe Gorath is correct. I don't know if the Nighthawks are in league with this Crawler, Delekhan, or both, or if we've just wandered into a conflict that has nothing to do with our mission, but either way it's to Delekhan's benefit. Which means we must help to end it.'

'How?' asked Owyn.

'Find out how this thing started, and see if we can figure out a way to get the two sides talking to one another. If the Earl can mediate the conflict, perhaps we can return this city to something close to order. Those lancers can only hold down the lid on this simmering pot so long; sooner or later someone's going to pull a sword or break a head, and a full-scale city riot will be under way.' He lowered his voice even more. 'And if most of the city's constabulary is on one side or the other, even those fifty lancers won't be able to stop it.'

Owyn nodded. 'What do you want us to do?'

Pointing to Gorath, he said, 'First light tomorrow, I'd like you up snooping outside the city. You know what to look for.' To Owyn he said, 'Do you know any of the prominent families of Romney?'

'Not well,' said Owyn, 'but as my father's a baron and I've got enough names to drop around, I should be able to get an invitation to tea or supper from someone around here.'

James said, 'Good. I'll snoop around.'

'Where?' asked Owyn.

James grinned. 'In parts of the city where wise men fear to go.'

Owyn nodded. 'What else?'

'Do you know a Baron Cavell, north of here?' asked James.

Owyn finished a mouthful of food. 'Corvallis of Cavell? I should. He's my uncle. My mother's uncle, actually, but only a few years older than her. Why?'

'Richard of Romney says he's being stalked by the Night-hawks.'

Owyn said, 'That doesn't surprise me. Uncle Corvallis always had a hot temper and an unforgiving nature. Made it easy for him to collect enemies. Still, I find it hard to imagine that anyone wants him dead.'

James shrugged. 'That's what Earl Richard said the Baron of Cavell claims.'

Gorath said, 'If they wanted him dead, he'd be dead.'

James said, 'Well, according to Richard, your uncle Corvallis is hiding out in a room in a house in the middle of Cavell Village, with armed guards in every room.'

Owyn nodded. 'The old keep was gutted mysteriously in a fire years ago. The family's been living in the best house in the village since then, and talking about restoring the old keep, but at this point it's still abandoned.'

James said, 'Well, we might have to go talk to your uncle if we can't find the Nighthawks down here.'

Gorath observed, 'I haven't noticed much difficulty in finding them.'

James nodded agreement. 'Too true.'

They finished their meal and turned in for the night.

The shout had barely registered on James the next morning and he was out of bed, grabbing his trousers and boots. Gorath was also awake and reaching for his sword. Owyn stirred on his pallet next to Gorath's and said, 'What?'

'Sounds like a riot is commenced,' said Gorath.

James listened to the sound and said, 'No, it's something else.'

He finished dressing and hurried down the hall to the stairs to the common room. As he approached the front of the building he could hear the voices from out in front. The landlord stood at the door to his inn, listening as people hurried by.

'What is it?' demanded James.

With a dark look, the innkeeper said, 'Murder. The cry is murder has been done in the night.'

'Murder?' asked Owyn, coming down the stairs. 'Who?'

'I don't know,' said the innkeeper. 'But they're saying it was done over at the Black Sheep Inn.'

James was through the door before the words had vanished from the air, Owyn and Gorath following. He didn't bother to go and saddle his horse, but rather sprinted through the streets, following the flood of people who swept along like a stream, heading across the bridge toward the main square of the city.

As he neared the square, he found a press of people being held back by a few men with pole arms, all wearing armbands. None of the Royal Lancers was in evidence.

James had to push his way through the crowd and when he reached the front, he was barred by a man holding a pike.

James pushed aside the pike shouting, 'On the business of the Crown!'

The man obviously wasn't prepared for that and hesitated, letting James, Gorath and Owyn pass. But he managed to keep others back as Richard, Earl of Romney, came striding across the square, toward the fountain. He saw James and exclaimed, 'Squire!'

James crossed to where he waited and said, 'My lord? What is it?'

Barely able to speak because of his rage, he pointed to the open door of the Black Sheep Inn and said, 'Look!'

James hurried to the entrance.

Entering the commons he saw Royal Lancers, sprawled across tables or on the floor, their eyes vacant and fixed. He needed no healer or priest to pronounce the men dead. He looked over at a cowering stableboy, who had found the bodies when he had come in for breakfast an hour earlier, and said, 'All of them?'

The boy was so terrified he could barely speak. 'Sir,' he nodded. 'The officer is in his room upstairs, and the sergeant and some of the others. The rest died down here.'

Gorath crossed to the table and picked up a mug of ale. He sniffed at it. 'Poison,' he said, 'or I'm a goblin. You can smell it.'

James took the mug and sniffed it, judging the moredhel's sense of smell keener than his own, for he could detect no odour beyond that of warm ale. He noticed a slight black sediment in the mug. He fished out a tiny bit with his finger, then touched it to the tip of his tongue. Spitting it out, he said, 'You may be right, and there may be poison in this ale, but what you're smelling is tarweed.'

'Tarweed?' asked Owyn, looking pale despite the number of corpses he had seen already.

James nodded, putting down the mug. 'Old trick in some of the seedier inns in the Kingdom. Tarweed is nasty stuff in large amounts, but in small doses it makes you thirsty. You lace bad ale with it, and the customers drink it like it was dwarven winter ale.'

'Can it kill you?' asked Owyn.

'No, but there are many tasteless poisons that can,' said James.

He turned to the boy and said, 'What's your name?'

'Jason,' the boy answered, terrified. 'What are they going to do to me?' he asked.

'Nothing, why?'

'I served these men, sir. My master always said the care of our guests was our responsibility.'

James said, 'Perhaps, but you couldn't know the ale was poisoned, could you?'

'No, but I knew something was odd, and I didn't say anything.'

James was now acutely interested. 'What was odd?'

'The men who came with the ale. We buy our ale from the Sign of the Upturned Keg down in Sloop. I know the waggon drivers. This time it was strange men.'

James took Jason by the shoulders and looked him in the eye. 'Is there anything you can tell us about these men, anything special?'

Jason stared at the ceiling a moment, as if struggling to remember. 'They were dark men, maybe Keshians, and they spoke oddly. And they seemed worried, but they didn't say anything. One wore a medallion that swung out from under his tunic when he leaned over to hand a keg down to his partner.'

'What did it look like?' demanded James.

'It had a bird on it.'

James glanced at Gorath and Owyn. 'What else?' asked James.

'They told me to forget I had ever seen them,' replied

Jason. 'And they smelled funny, like sailors from Silden do when they come here, like sun on canvas and flowers.'

Gorath and Owyn began inspecting the room, while James went outside. He saw Earl Richard, rooted to the same spot he had occupied when James had entered the inn. The shock of the murders had rendered the Earl nearly unable to function. James had seen it before with men who were unused to bloodshed. He hurried to the Earl and said, 'My lord, what do you propose?'

Blinking as if he had difficulty understanding James, Richard echoed, 'Propose?'

James pointed at the crowd and said, 'You must tell them something. Disperse them before things get any uglier than before. Then the bodies must be attended to.'

'Yes,' said Earl Richard. 'That's so.' He mounted the fountain and stood where everyone could see him. 'Citizens of Romney,' he shouted, and as the words came from him, James could see that speaking before the citizenry was something the Earl did often, for the familiarity of the task returned his wits.

'Go to your homes!' commanded the Earl. 'Stay calm. Black murder has been done and those responsible will be hunted down and punished.' He jumped down and waved over a constable. 'I want someone from the Riverpullers and the Ironmongers here in five minutes.'

'Damn!' said the Earl to James. 'I need send to Cheam for more troops. Black Guy won't be pleased when he learns fifty of the King's Own have died in my city.'

'Nor will the King,' observed James. Seeing the Earl's face cloud over at the mention of King Lyam, James said, 'My companions and I will do anything we can to help.'

'The best thing you can do right now, squire, is find out who is behind this.'

'I already know,' said James. He told them of the tarweed and the two men who appeared to be from Silden.

'Nighthawks!' whispered the Earl, so as not to be over-heard by any of the crowd who were slowly leaving the area. 'Damn! I almost wish it had been Damon Reeves or Arle Steelsoul behind this.'

'Why them?' asked James.

'Because then I could hang one or the other with cause and end two problems for the price of one. Reeves runs the Riverpullers, and Arle Steelsoul is the head of the Ironmongers' Guild. They are at the heart of the dispute.' He indicated two men approaching. When they were standing before the Earl, he said, 'Tell your respective factions that I have had enough with violence in Romney. I hold the head of the Riverpullers and Ironmongers personally responsible for the good behaviour of both sides of this dispute. Any further violence and I will hang them, side by side, from the city gate. Carry word back to them now!'

The first man, one of the Ironmongers, said, 'But Arle Steelsoul's down in Sloop!'

'Then carry word to Sloop,' instructed the Earl.

James said, 'M'lord, I will do that.' The two men exchanged glances, as if asking who the stranger was to bear such tidings to the leaders of the two warring factions.

The Earl said, 'Pass the word that Arle and Damon's lives will be forfeit if there are any more problems in my city.' The two men bowed and ran off.

'Can you enforce the threat, m'lord?' asked James when the men were out of earshot.

'Probably not, but it may shock them into behaving them-selves until the next detachment of soldiers arrives.' He looked at James. 'Why do you choose to go to Sloop?'

'Because that's where the poisoned ale is from, and because I think we need to continue on down to Silden after that.'

'Then tell Steelsoul and Michael Waylander that I expect both men to be here in three days' time, along with Reeves and the other local leaders of the various factions, and should either not appear, I will know he has a hand in black murder. I'll issue the death warrant myself. If they both show up, I'm locking all of them in a room and neither side will be permitted to leave until we have a settlement of these differences. I don't care if they have to pee on the floor, or die of starvation, I'll have an end to this business before any of them sees the sun again.'

Convinced of the Earl's earnestness in the matter, James said, 'My companions and I will be off in an hour, m'lord.' He bowed and returned to the Black Sheep, where two workers were helping Jason move the bodies so they could be piled up on a waggon and taken from the city for cremation. Owyn waved James over.

'Find anything interesting?'

'Just this,' said Owyn. He held out two items. One was a small silver brooch, looking like an oversize spider.

'What's this?' asked James.

'Turn it over,' said Gorath.

James did as he was bid and saw a large groove running down the centre of the item. In it a tightly-packed gummy substance could be seen. James lifted the device to his nose and sniffed. 'Silverthorn!' he said.

'Are you sure?' asked Owyn.

'I'd recognize that odour anywhere, trust me,' replied James.

'It's an assassin's tool,' said Gorath. 'You run the edge of a dagger along that groove and even if you don't strike a killing blow, the victim dies within hours.'

'What else?' asked James.

Owyn held out a brass tube with glass at each end. 'A spyglass?' asked James.

'Look through it,' suggested Owyn.

James did and his perspective altered. The colours through the glass changed and he suddenly saw shifting patterns on the clothing of his companions as well as on the walls of the building. Pulling it away from his eye, he said, 'What is this?'

Owyn said, 'It's magic. I will have to study it, but I think it lets you see things you otherwise can't see, such as magically-hidden items.'

James looked down at the two items. He wished he had better clues, but these two would have to do as a start.

EIGHT

Secrets

Dark shapes moved in the evening shadows.

James pointed to them and Owyn asked, 'What?'

Gorath said, 'I see them.'

They had ridden south at midday, pushing the horses as much as possible, to reach the village of Sloop and deliver the Earl's ultimatum to Steelsoul and Waylander. As sundown approached, they had crested a rise and come within sight of town. Armed men were filtering through the trees at the north end of the town, heading toward a clump of houses.

Gorath urged his horse forward, pulling his sword. James and Owyn were on his heels a moment later. They charged the men, while James started shouting, 'Alarm! Raiders in the village!'

He knew that depending on the make-up of this village, the response to a call of alarm would either be for the men of the village to rush out with weapons in hand, or for doors and windows to be locked down. In the west he knew there would be a dozen men in the streets to meet the invaders in a minute. Here in the relatively calm east, he wasn't so sure.

As they passed the first house, he saw a curious face peeking through a window. Again he shouted, 'Raiders in the village! To arms!'

The man slammed shutters, and James could imagine him barring the door as James left the house behind.

Gorath was upon the first swordsman, leaping from his horse onto the man. James considered that he probably should devote at least one afternoon teaching the dark elf how to fight effectively from horseback.

Owyn, on the other hand, had become quite adept at using his heavy staff from horseback, cracking skulls and breaking arms with quick efficiency.

Within minutes the raiders were on the run, heading back into the woods. James rode to where Gorath seemed poised to give chase and shouted for him to halt. 'It'll be dark soon,' he said. 'Even with your woodland skill, we don't want to try chasing a half-dozen angry Nighthawks into a dark forest.'

Gorath said, 'Agreed,' and turned to find his horse.

James went to the house that was the obvious target of the raid and dismounted. He pounded on the door. 'Open in the King's name!' he shouted.

Through a viewing slit a pair of eyes, wide with fear, regarded him. The door opened and Michael Waylander said, 'Squire. What was all that noise about?'

James said, 'It looks like someone is taking the game to a higher stake. We just chased off a band of Nighthawks coming to see you.'

Waylander turned pale. 'Nighthawks?' His knees went weak and he gripped the doorjamb to stay on his feet. 'What have I got myself into?'

James said, 'That's what we've come to talk about.'

Gorath and Owyn tied their horses next to James's and came to the door as Waylander stepped aside to admit them. It was a modest house, but James noticed at once it was well kept. There was enough wealth evident in the furnishings and appointments that it was clear Michael Waylander was very well situated for a common worker in a small village. The house, while not large, had three rooms, a bedroom visible through a door, and James saw the bed was a well-carved four-poster with a mesh netting and canopy. Through the other door James could see a kitchen. Waylander sat heavily on a chair, and James sat in the other one next to a table.

'Someone wants you dead, Michael,' said James. 'Who could that be?'

Waylander sat back, a look of defeat on his face. 'I'm a dead man.'

'Maybe not,' said James. 'I represent Prince Arutha and while you've obviously irritated some powerful people, the Prince of Krondor is still the most powerful man in this nation after the King. If you co-operate, I may be able to get you under his protection.'

Waylander stared off into space a moment, as if thinking. 'I'm in over my head. I'll do whatever I must to get out of this.'

James leaned forward and suggested, 'Why don't you start with what "this" is.'

'About a year ago, some men came to me from Silden. They had an idea, and I took that idea to Arle Steelsoul.'

'What was the idea?'

'The idea was to take control of all the business along the river, from Silden to the small villages in the mountains.'

'How were they to accomplish this?' asked James.

'They said they had connections in the Riverpullers, who had told them the Guild was going to raise prices for hauling cargo up the river.'

'So the Guild wanted to raise their rates?'

'Yes,' said Waylander. 'They're usually cautious about that, because if the rates go too high, merchants start using waggons to send goods north along the King's Highway.'

'But if there was a lot of trouble on the Highway, merchants would be forced to use the barges and the Riverpullers,' finished James.

'Yes.' Waylander nodded agreement. 'These men said that they could ensure the Riverpullers would have no competition. Then we, Arle Steelsoul and I, would organize the other guilds in Romney and the surrounding villages to stand against the Riverpullers. When things got bad enough, the King would declare martial law, and the Riverpullers would be put out of business.'

'And what does it matter if some heads get broken along the way?' asked Owyn dryly.

'Waylander,' asked James, 'what made you think the Riverpullers would be out of business if the King declared martial law?'

'We planned on having Damon Reeves, head of the Riverpullers' Guild, murdered.' He hung his head as if ashamed at this admission. 'I didn't want that, but by the time they told me of the plan, I was in too deep. They said they'd make it look like Nighthawks did it, so that no blame would fall to us. In fact, they said they'd make it look like someone within the Guild did it, to get Reeves out of the way, and the Guild would fall apart from dissension within. I've known Damon for years; he's an old friend, but there was nothing I could do.'

James glanced at Gorath and Owyn. 'Whose idea was it to cast blame on the Nighthawks?'

'The men from Silden,' said Waylander. 'Why?'

'Just that the notion is familiar to us.'

Owyn realized James was talking about the false Nighthawks in the sewers of Krondor and nodded in understanding.

'What should I do?' asked Waylander.

'Get Steelsoul, get to Romney, and sit down with the Riverpullers and make peace. If you don't, the Earl will hang you two and Reeves, and start over with whoever replaces you.'

'The Earl's never resorted to threats before. Why is he suddenly threatening us now?' asked Waylander.

'Because someone just murdered fifty Royal Lancers in his city,' answered James.

Waylander's eyes widened and his face turned ashen. 'Fifty! Gods of mercy!' He gripped the table and said, 'Who could do such a thing?'

'Chance has you crossing paths with the Nighthawks, it

seems,' suggested James. 'And by all appearances they don't seem all that pleased by these attempts at implicating them in deeds for which they are not responsible. No matter how clever you gentlemen thought you were being, you were being played for fools by agents of a man who is called "the Crawler". He's attempting to dislodge the Mockers in Krondor and seems to want to control the docks in the eastern cities as well. They were not helping you; you were being set up to create a situation where they would emerge in control after you, Reeves, Steelsoul, and anyone else inconvenient to their goals were out of the way. It wouldn't surprise me if the Crawler's agents hadn't leaked the information to the Nighthawks about your attempting to hang the blame for Reeves' murder on them.'

'As if another charge of murder is going to make them any more hunted,' Gorath observed.

'True,' said James, 'but it's been my experience that criminals take a certain pride in their own crime, but want nothing to do with blame for crimes for which they are not responsible. It's odd, I know, but that's the way it is.'

'You talk as if you've known a lot of criminals,' said Waylander.

'Yes, I do, don't I?' James's smile lacked even a suggestion of warmth.

'What do I do after I see the Earl?'

'I suggest you beg for leniency,' said Owyn.

James nodded. 'People have died as a result of your choices, and you and Steelsoul have much to answer for. But if you help the Earl restore order and help us uncover those behind this plot, we'll do what we can to keep you off the gibbet.'

'Maybe I should just run,' said Waylander.

'You won't reach Silden,' said James. 'They would be on you like hounds on a hare, and where would you go, anyway?'

'I have connections in Kesh,' said Waylander. 'If I can get

to Pointer's Head, I can take a caravan over the Peaks of Tranquillity.'

'Well, don't do anything rash,' said James. 'If my friends and I have our way, the Nighthawks will not be a problem much longer. My advice is to see the Earl, then sit tight. I'll get word to you when it's safe.'

'But what about the men in Silden?'

James stood up. 'They're also a problem.'

'But I only know them by sight and first names, Jacob, Linsey and Franklin, and they may not even be their true names.'

'Probably not,' said James. He took the spyglass and the silver spider out of his travel bag and said, 'What can you tell me about these?'

Waylander said, 'The spider I got from a trader named Abuk. He travels the roads between Malac's Cross and here, stopping in at Silden each way. I last saw him there, so he may be on his way toward us right now. He drives a trader's waggon painted green, with his name in red letters on the side.'

Owyn winced at the description. 'We can hardly miss that.'

James's expression turned dark. 'We found this spider this morning among the bodies of the dead lancers.'

Waylander said, 'It can't be the same one, then!'

'Why?' demanded James.

'I bought one from Abuk, but I gave ours to the false Nighthawks who were sent to kill Damon Reeves.'

James looked at the device and said, 'There may be more than one, but you'll need more proof of your innocence than that.'

Waylander examined the spider, then said, 'Look!' He pointed to the groove containing the poison. 'I don't know what this is, but mine had deadly nightshade in it!'

Gorath said, 'Silverthorn would be hard to locate this far south.'

'But not impossible,' said James. 'Still, I'm inclined to believe you. What about the spyglass?'

'I don't know anything about that,' said Waylander, 'but it's the sort of thing Abuk trades for as well.'

James led the others to the door. 'Get to the Earl, Michael,' he said. 'You and Arle should be there before sundown tomorrow if you value your heads. We're in the inn until dawn, and then we're going south.'

'I'll walk with you as far as Arle's house,' said Waylander. 'And then we'll see the Earl tomorrow. Where south are you going?'

'First to Silden to find Abuk and those three men you mentioned. If we have any luck, we'll put paid to this mess within a few days.' Waylander said nothing, and James knew it was because even if all the Nighthawks and Crawler's men vanished overnight, there would still be crimes to pay for. But even years in a dungeon, thought James, were better than dying. At least in a dungeon there was the chance of escape.

The last thought made him smile as he headed up the road toward the inn.

As they neared the town of Silden, they slowed. A band of men were also riding toward the town, coming in from the west. 'We don't know they're looking for us,' said James. 'But as many times as you've been attacked, Gorath, I'd just as soon wait to see what they're up to.'

Gorath had no disagreement, so he remained silent. The riders crossed over the bridge which arched over the River Rom into the town proper. Because it was built on a bluff that sloped down to a deep harbour, Silden had no foulbourgh outside the city walls. Rather, a series of small villages dotted the coastline around the bay of Silden, and a large village dominated the western shore of the bay, on the other side of the bridge.

They rode into the northern gate of the city, and passed a

bored-looking pair of city watchmen. James turned to Owyn and asked, 'Any friend or relatives here?'

'Not that I'm aware of,' said Owyn. 'Or at least none my father would admit to.'

James laughed. 'I can understand that. This isn't exactly a garden spot, is it?'

Silden was only important to two groups: those who lived in it and smugglers. The majority of trade coming up the river to the north entered through the much larger trading port of Cheam, which had spacious docks, a huge warehouse district, and was the second largest port on the north shore of the Kingdom Sea after Bas-Tyra. Silden was therefore a far more profitable destination for those seeking to conduct business without benefit of Kingdom Customs officers. They made an attempt to curtail smuggling, but with the host of villages within a day's ride to the east and west, keeping smuggling under control was impossible. As a result, control of Silden had for years been an ongoing goal of competing criminal gangs, from the Mockers of Krondor, Keshian drug smugglers, and bully gangs from Rillanon, to an alliance of local thieves. This constant struggle had turned Silden into the closest thing to an open city seen in the Eastern Realm of the Kingdom.

The Earldom of Silden, while a reasonably attractive fiefdom, with rents and income sufficient to keep a noble family in style, was an absentee office. The last Earl of Silden had died during the Riftwar, in the great attack by King Rodric IV against the Tsurani in the final year of the war. King Lyam had yet to award the Earldom to anyone, which was fine with the Duke of Cheam, who presently enjoyed the income from the property in the Earldom. James was of the opinion it should be turned into a proper duchy and run from here in the city. A resident noble would clear up a lot of the problems of this valuable port city. He would have to mention it to the Prince when he returned, but for the moment, it was still a neglected, backwater town without proper over-sight.

The upshot of this situation was an almost complete absence of law and order in Silden, beyond that which was enforced by the local constabulary. And from what James could tell, it ended where the market district of the city turned into the waterfront, and at a boulevard marked by a sign of four gulls in flight. One side of the street was marked by prosperous-looking shops and homes, the other by inns and warehouses. Down the middle of the street a long red line had been painted.

'What is that?' asked Gorath as they rode across to it.

'A deadline,' said James. 'If you're brawling over there, no one cares. Brawl on this side, and you're off to the work gangs.'

He motioned for them to cross the deadline and as they entered the dock district, he said, 'Ah, I love a town where they let you know how things stand with no apology.'

Gorath looked at Owyn and shrugged. Then he asked, 'Why is it called a deadline?'

Owyn said, 'In the past if you were caught after curfew on the wrong side by the soldiers of the King, you were hanged.'

They rode through a series of dark streets, bounded on either side by high warehouses, and crossed another fairly large street, rumbling with waggons and large men pushing carts piled high with goods. Then they were looking at the harbour below, a jumble of docks and jetties, some stone, mostly wood, pushed hard against one another. Small boats were moving in and out of the harbour. Silden was blessed with one saving grace, the high bluffs upon which the three riders now stood, which provided shelter from the harshest winter storms.

James conducted them down the long roadway which led to the docks and pointed to an inn in front of which hung a sign made from an old ship's anchor, painted white. A modest stabling yard stood to the side and when James rode in, a grubby-looking boy hurried over. 'Pick their feet, give them hay and water, and rub them down,' said James as he dismounted.

The boy nodded and James said, 'And tell whoever's interested that I would consider it a personal courtesy if these animals were here in the morning.' He made a small gesture with his thumb and the boy nodded slightly.

'What was that?' asked Owyn.

As they entered the Anchorhead Inn, James said, 'Just a word dropped in the proper ear.'

'I mean the thing with the thumb and fingers.'

'That's what let the boy know I deserved to be listened to.'

The common room was seedy and dark, and James looked around at its clientele. Sailors and dockhands, soldiers of fortune looking for an outward-bound ship, ladies of nego-tiable virtue, and the usual assortment of thugs and thieves. James took them to a table in the rear and said, 'Now we watch.'

'For what?' asked Gorath.

'For the right person to show up.'

'How long do we wait?' asked Owyn.

'In this hole? A day, two at the outside.'

Gorath shook his head. 'You humans live like . . . animals.'

'It's not so bad once you've got used to it, Gorath,' said James. 'It's a fair improvement over some places I've called home.'

Gorath said, 'That is an odd claim for one who serves a prince of his race.'

'Agreed,' conceded the squire, 'but none the less true for being strange. I have had an unusual opportunity to improve my situation.'

'The opposite is my fate,' said Gorath. 'I was a clan chieftain; I was sought out in council and was counted among the leaders of my people. Now I am sitting in squalor with the enemy of my race.'

James said, 'I am no one's enemy lest he harm me or mine first.'

Gorath said, 'I can believe that, squire, though it strains my senses to hear myself saying it; yet I can't say that for most of your race.'

James said, 'I never claimed to speak on behalf of most of my race. If you've noticed, we're often a great deal more busy killing one another than we are causing problems for the nations of the north.'

Suddenly Gorath laughed. Both Owyn and James were startled by the sound, surprisingly musical and full. 'What's so funny?' asked Owyn.

Gorath's smile faded and he said, 'Just the thought that if you were a little more efficient killing one another, I wouldn't have to worry about a murderous dog like Delekhan.'

At mention of the would-be conqueror, James was reminded of the importance of unravelling the knotted cord of who was behind which plot. So far he had decided that this Crawler, whoever he might be, was more a problem for the Upright Man and his Mockers, and Prince Arutha, and whatever other local nobles he was plaguing, but his part in Delekhan's plans was coincidence, not design.

The Nighthawks were obviously working with either the Crawler, the moredhel, or both. And what caused James to worry was that they might be again the pawns of the Pantathian Serpent Priests. At some point James would bring up the serpents with Gorath, but not here in this public a place.

The barmaid, a stout woman who had probably been a whore in her youth, but now could not rely on her faded looks to earn her livelihood, came over and with a suspicious look at Gorath asked their pleasure. James ordered ale, and she left. James returned to his musing.

There was another player in this, some faction who was orchestrating all this turmoil in the Kingdom, either the Pantathians or someone else and that was what had James concerned. Going over what Gorath had told Arutha and James several times, he said, 'I would give a great deal to know more about those you call the Six.'

Gorath said, 'Little is known of them, save by Delekhan's closest advisors, and I know of no one who has actually met them. They are powerful, and have provided my people with weapons in abundance. But Delekhan's enemies have been disappearing suddenly. I was called to council and taken on the road to Sar-Sargoth and locked away in the dungeon by Narab, Delekhan's chief advisor.'

James said, 'You didn't mention that part before.'

'You didn't ask about what I had been doing before I met Locklear,' said Gorath.

'How did you escape?'

'Someone arranged it,' said Gorath. 'I'm not sure who, but I suspect it was an old . . . ally. She is a woman of some influence and power.'

James was suddenly interested. 'She must have a great deal of influence to get you free right under Delekhan's nose.'

'There are many close to Delekhan who will not openly oppose him but would be pleased if he failed; Narab and his brother are among them, but as long as the Six serve Delekhan, they will as well. Should anything befall Delekhan before he consolidates the tribes, any alliance he has forged will disintegrate. Even his wife and son are not fully trusted by him, and for good reason. His wife is Chieftain of the Hamandien, the Snow Leopards, one of the most powerful clans after Delekhan's own; and his son has ambitions that are obvious.'

Owyn said, 'Sounds like a happy family.'

Gorath chuckled at that, his tone ironic. 'My people rarely trust those who are not of our own family, tribe or clan.

Beyond that are political alliances and they are sometimes as fugitive as dreams. We are not a trusting people by nature.'

'So I have determined,' said James. 'Then, for the most part, neither are we.' He slowly stood up. 'Excuse me. I'll be back in a moment.'

He passed the barmaid who ignored him as she brought the ale to the table, which forced Owyn with ill humour to pay for the drinks from his meagre purse. Gorath found this amusing.

James crossed to where a man had emerged from the back room, dark skin and beard marking him as one of Keshian ancestry. 'Can I help you?' he asked with an appraising look. By his accent, he was a Keshian by birth. He was thin, and James assumed dangerous, and while his close-cropped beard was greying, he was probably still vigorous enough to be a deadly opponent.

James said, 'You're the owner of this establishment?'

'I am,' he said. 'I am Joftaz.'

Lowering his voice, James said, 'I am here representing interests that are concerned with some downturns in their business of late. There are difficulties stemming from the activities of men who have been most recently both up in Romney, and to the west.'

Joftaz regarded James with an appraising eye. 'Why mention this to me?'

'You live in a place where many pass through. I thought perhaps you might have heard something or seen someone.'

Joftaz laughed in a jovial manner that was entirely unconvincing. 'My friend, in my line of work, given where we are, it is in my interest to hear nothing, notice no one, and say little.'

James studied the man a moment. 'Certain information would have value.'

'How much value?'

'It would depend on the information.'

Joftaz looked around and said, 'The wrong thing said in the wrong ear could end a man's life.'

'Daggers have points,' said James, 'and so do you.'

'On the other hand, I do find myself in need of some help in a delicate matter, and for the right man I could possibly remember a few things I've heard or faces I've seen.'

James nodded. 'Would this delicate matter be aided by a sum of gold?'

Joftaz smiled. 'I like your thinking, young man. What may I call you?'

'You may call me James.'

For an instant the man's eyes flickered and he said, 'And you are from . . . ?'

'Most recently, the village of Sloop, and before that Romney.'

'Then the men you seek who had been recently in Romney are involved in some matter up there?'

'Some matter, but before we discuss what I need to know, I need to know the price.'

Joftaz said, 'Then, my young friend, we are at something of an impasse, for to tell you any of my need is to tell you all my need, and as they say, "in for a copper, in for a gold".'

James smiled and said, 'I'm hurt, Joftaz. What must I do to win your trust?'

'Tell me why you seek these men.'

'I seek them as nothing more than a link in a chain. They may lead me to another, one with whom I have some serious issues. He is one behind murder and treason, and I will have him to the hangman or dead at my feet; either is fine with me.'

'You're the King's man, then?'

'Not directly, but we both respect my employer.'

'Then swear by Ban-ath you will not betray me, and we shall strike a bargain.'

James's grin broadened. 'Why by the God of Thieves?'

'Who better? For a pair of thieves such as we.'

'By Ban-ath, then,' said James. 'What is your need?'

'I need you to steal something from the most dangerous man in Silden, my friend. If you can do that, I will help you find the men for whom you are looking. Assuming you survive, of course.'

James blinked. 'Me, steal? Why would you think I would steal for you?'

'I have lived enough years to know where eggs come from, young man.' He smiled. 'If you are willing to swear by Ban-ath, you've walked the dodgy path before.'

James sighed. 'I would be forswearing my oath to speak truly if I denied such.'

'Good: to the heart of the matter then. There is just a short walk from here a house, in which dwells a man, by name Jacob Ishandar.'

'A Keshian?'

'There are many from Kesh who reside here.' He touched himself on the chest. 'Such as I. But this man and others like him have but recently come to Silden, less than two or three years ago. They work on behalf of one who is a spider, sitting at the heart of a vast web, and like the spider, he senses any vibration along that web.'

James nodded. 'You speak of one known as the Crawler?'

Joftaz inclined his head, indicating that this was the case. 'This was never what one might call a peaceful community, but it was orderly after a fashion. With the Crawler's men – Jacob and two called Linsey, and Franklin – came bloodshed and pain beyond what is reasonable for men in our line of work to endure.'

'What of the local thieves, and those with ties to Rillanon and Krondor?'

'All gone, save myself. Some have fled, others . . . disappeared. Any thief I contacted in Silden today would be working for the Crawler. Being Keshian by birth, I think these men did not recognize me for one such as those they sought to destroy. There are still a few of us in Silden who survived, but we conduct no business except what we do in the open, such as my inn. Should these interlopers' enterprises fail, there will be enough of us returning here to reclaim what was taken from us.'

James scratched his chin as he thought. 'Before I agree, let me show you something.' He produced the silver spider. 'Do you know this?'

'I have seen such before,' he said. 'They are rare and when one comes my way I take notice. They are crafted by a smith in a village in the Peaks of Tranquillity. Those that reach the Kingdom come from Pointer's Head or Mallow Haven.' He took it from James's hand and inspected it. 'I've seen bad copies, but these are far finer. You can't work silver like this and have it endure unless you have the knack.'

'Odd sort of bird buys an item like this.'

Joftaz smiled. 'Night birds, for the most part. You play a dangerous game, my friend. You are just the man I seek.'

'Well, then, can you tell me who you sold this one to?'

'Yes, I can, and more.' Joftaz lost his smile. 'But not until you conduct some business for me.'

'Then to specifics.'

'This man I mentioned, Jacob Ishandar, is chief among those recently come from Kesh. He has in his possession a bag – ' he held his hands apart, indicating a bag the size of a large coin purse or belt pouch ' – and the contents of that bag are worth enough to underwrite his operation here in Silden for the next year.'

'And you want me to steal that bag?'

Joftaz nodded.

'I would think you able to undertake such a task yourself,' said James.

'Perhaps, but I must continue to live here in Silden, success or failure. Should you fail, I will still be here.'

'I see. What's in the bag.'

'Heart of Joy,' said Joftaz.

James closed his eyes a moment. Joy was a common drug

in the poor quarters of most cities in Kesh, and showed up from time to time in Krondor and other port cities in the Kingdom. A small amount consumed in wine or water would induce a pleasant euphoria for up to a night. A slightly larger dose would transport the user to a state of happiness that could last days. If the dose was too large, the user would be rendered unconscious.

Heart of Joy was a different thing. It was the essence of the drug, compounded in such a way as to make it easy to transport. When sold, it would be mixed in with a harmless powder, often powdered sugar or even flour, anything that would dissolve. By weight it was worth a thousand times more than Joy when sold on the streets of the city.

'A bag that size is worth – '

'Enough to ensure that Jacob will have to run for his life when the Crawler finds out, and any who might be held responsible as well – say Linsey and Franklin – will flee along with him.'

James filled in, 'Leaving a void into which you can step to re-establish business locally in a fashion more to your liking.' Narrowing his eyes, James added, 'And he who finds it will find anxious buyers willing to say nothing about where the drug came from, realizing enormous profits.'

With a smile, Joftaz said, 'Well, there is that.'

'So, if I get that bag, you put the Crawler's agents in Silden out of business and make yourself a fortune in the process.'

'If all goes well.'

James said, 'We'll be in the corner, my friends and I. When you are ready, tell me where I must go and what I must know.'

'We close the common room at midnight. Wait until I do, then we shall see about your needs.'

James returned to the table, and Owyn said, 'What did you find out?'

'That nothing in life is ever free,' said James, sitting down and leaning his chair back against the wall, settling in for a long afternoon's wait.

The house was apparently deserted, its occupant away on some errand. Gorath was instructed to stand a few doors down, watching for anyone coming up from the docks. Owyn stood on the other side of the street, watching in the other direction. Both agreed to co-operate, both expressing their doubts as to the wisdom of this enterprise.

James quickly inspected the door for obvious alarms and found none. He judged the lock an easy enough one to pick, but just for reassurance, he ran his thumb along the doorjamb. Unexpectedly he found a crack in the wood, which moved under his thumb. Carefully he pushed on it, and heard a slight click from within. Pushing harder, he moved the wood. From behind it protruded a piece of metal.

James removed a brass key from a hiding place in the wood. He almost laughed. It was an old, very simple trick, and served two purposes: the key was never lost if the owner was in a hurry leaving someplace else, and it disarmed whatever trap waited inside. In the daylight, James expected he could have looked for hours and not seen it, but an old thief had once taught him to trust his other senses, including touch. Running the thumb over the doorjamb occasionally brought splinters as its only reward, but the sound of that click made the hours James had spent fishing splinters out of his thumb with a steel needle worth it.

James still knelt as he pushed the door open slightly, ready for anything that would alert him to another trap. By kneeling, any crossbow bolt aimed at the door should fly overhead.

The door slid open easily and no device sent death his way. He moved quickly through the door and closed it behind him. He inspected the room without moving. He never knew where someone would hide valuables, but most people were predictable. This time, however, he considered the owner of this place was not 'most people', but someone who would do something unpredictable. So his first choice was to look for something out of place.

The room was undistinguished. A simple table, a large breakfront clothes closet, and a bed. A door to a rear yard where the outhouse would be. A fireplace, above which rested potted plants on a wide mantel, and next to that a door leading into a small kitchen.

Then it registered on James. Potted plants? He moved to inspect them. They were dry and dying, and he knew the reason why. He couldn't remember the name of the variety, but Princess Anita had struggled to raise the same plants in her garden in Krondor. She had remarked that they were difficult to grow in soil with as much salt as the soil near the palace, and that they demanded a great deal of sunlight.

Silently, James asked, why would a leader of a gang of cutthroats in a pesthole like Silden have potted plants on his mantel? He carefully lifted the pots, one at a time, until he picked up the one on the far right. It was lighter than the rest. He lifted the plant and it came away, devoid of dirt on the roots. Under it he found a bag, and he returned the plant to the pot and opened the bag. In the dim light coming from the sole window to the house he saw what he expected to see, a slightly yellowish powder.

He tied the bag and moved quickly to the door. One backward glance reassured him he hadn't inadvertently touched anything. He slipped through the door and closed it behind him. He locked it, and returned the key, resetting whatever trap had awaited the unwary on the other side.

He motioned without looking at either of his friends and they returned to the Anchorhead Inn. As they neared the door at the rear, left open for them by Joftaz, James felt a flush of excitement. No matter how high he might someday rise in the King's service, there was a part of him that would always be Jimmy the Hand.

Inside he handed over the bag to Joftaz and said, 'Well, then, your part of the bargain.'

Joftaz admired the bag of powder for a moment, then put it behind the bar. 'To find the owner of that spider, you must seek out the trader, Abuk. I have sold four such as this to him over the last two years.'

James produced the spyglass. 'What about this?'

Joftaz admired the glass and held it up to his eye. His eye widened and he put down the glass, glancing around the room. 'This is a dangerous thing, my friend.'

'Why?'

'It shows secrets, and some secrets are worth killing to preserve or to learn.' He handed the spyglass back to James. 'I have heard of such as these. They are modest-looking, but valuable. You pierce illusions, see traps and hiding places with a glass like that. I have heard of such glass being fashioned for generals to pierce the fog and smoke on the battlefield.'

'Do you know who might have sold this?'

'Again I say, Abuk. Had this item come to you from any other source, I would not guess, but if you found it near the spider, I suspect they were both sold by him, and to the same man.'

'Then we need a room for the night, my new old friend, and then we're off in search of Abuk.'

They shook hands and Joftaz said, 'You serve your king well, my new old friend, for not only do you seek out Nighthawks who do black murder in the darkest hour of the night, you have rid Silden of the plague of the Crawler. Jacob and his companions will be on the first ship bound for distant lands once word of this reaches their employers. Now, I'll show you to your rooms, then I must find a certain rumour-monger to spread word that three Keshian gentlemen now residing in Silden have just sold a great deal of Heart of Joy to a smuggler bound for the island Kingdom of Roldem.'

Joftaz took them up to a room and bid them goodnight, and informed them that they should expect to encounter Abuk on the road between Silden and Lyton, as he was due back from there in the next few days. James settled in and quickly fell asleep, feeling at last he was making some progress in unravelling these mysteries.

NINE

Suspect

The mules lumbered up the road.

There was no mistaking the waggon as it hove into sight around a bend, a day's ride east of Silden. The green waggon had huge red letters on the side, proclaiming 'Abuk. Trader in fine wares.' The driver was a large, bull-necked man with an impressive mane of flaming red hair and a long beard that reached to his belt. If a dwarf could grow to more than six feet in height, this is what he'd look like, thought James as they halted before the waggon.

'You're the trader, Abuk?' asked James loudly.

The trader reined in his team of mules. 'It's what is written in large letters on the side of this waggon, stranger, so either you can't read or you're oblivious to the obvious. I am Abuk.'

James grimaced at the remark about the obvious. 'Well, you could have stolen his rig.'

'True, and I could have cut his hair and beard to create my disguise, as well. But I didn't.' He regarded the three riders before him. 'What may I do for you?'

'We are in the market for some information.'

Abuk said, 'Information is often my most profitable commodity.'

James walked his horse close enough to the buckboard of the waggon to hand over the silver spider. 'Can you tell me to whom you sold this?'

'Yes,' said Abuk. 'For the sum of a hundred golden sovereigns, I can.'

James grinned, and there was nothing but menace in his

smile. 'Or we could arrange for you to have a discussion with the Royal Interrogator regarding your part in the death of fifty of the King's Own Royal Lancers.'

'What?' demanded the startled Abuk. 'Fifty Royal Lancers were murdered?'

'In Romney,' supplied Owyn.

The trader was silent for a moment, calculating his chances of survival against his potential for profit, if James was any sort of judge of men. Finally he said, 'I take no responsibility for that act; I merely sell goods which are not banned by law.' He handed the spider back to James. 'This is one of two I sold in the north. A poor imitation was sold to a man named Michael Waylander in the village of Sloop. He is a prominent member of the Glaziers' Guild in the City of Romney. The other was sold to a man whose name I do not know, but I know he is from the north.'

James showed Abuk the spyglass. 'What of this?'

'You have proven the man you seek is the one I described, for he also purchased this glass. I sold both items to him at the Queen's Row Tavern in Malac's Cross, and you might inquire there of the innkeeper, who seemed to know this man. He was an exceptional chess player, by what I overheard.'

'If you met him in Malac's Cross, why then did you say he was from the north?'

'Because I overheard the innkeeper ask him if he was returning to the north, and the man said he was indeed heading home.'

James did not look pleased. 'We must then return to Malac's Cross.'

Abuk said, 'I might be able to save you a journey, for a small fee.'

James asked, 'How small?'

'A dozen golden sovereigns, I think.'

'Five, I think, and I forget your name when I speak to the King's Inquisitor.'

'Done,' said Abuk.

James gave him the money and the man said, 'Now that I recall, he did mention the town of Kenting Rush.'

James looked at Owyn, who nodded. 'I know it. It's north of my Uncle Corvallis's home in Cavell Village.'

Abuk looked at Owyn. 'Your uncle is the Baron Corvallis?'

Owyn said, 'Yes, he is.'

'I know him,' said Abuk. 'He's a man of ill humour, if you don't mind me saying so.'

Owyn grinned. 'No one who knows him will argue that.'

'If we are done?' asked Abuk to James. James indicated they were, and the vivid green waggon started forward again.

After Abuk was safely away, James turned to Owyn. 'What do you think? Malac's Cross or north to Kenting Rush?'

Owyn said, 'Kenting Rush is a small town, barely more than a dozen shops and inns. Mostly farmers and small estates in the area. There can't be too many men matching the description of the man we seek in residence there.'

Gorath said, 'Good, because time is growing short. It's been more than a month since I left my homeland and Delekhan's power grows while we seek out information. It would do us no good to discover his plans by witnessing them executed.'

'A good point,' said James, turning his horse around. 'Let us head north.' He urged his mount forward and set off at a brisk trot. A few minutes later they overtook and passed Abuk, and with a wave of farewell, continued down the road.

The passage between their encounter with Abuk and the turn-off to the City of Romney went without a hitch. They paused in Romney to change horses and see if things were calming down there.

Michael Waylander, Damon Reeves and Arle Steelsoul

had heeded the Earl's warning and appeared within days of the message being delivered. They were now locked in earnest negotiations with the other guild leaders to end the struggle between the rival guilds in the city and order was slowly returning to Romney.

The next morning, James, Gorath and Owyn departed on fresh horses, and hurried north through the rolling farmland that bordered the River Rom. The towns and villages along the river were undistinguished, much like the village of Sloop, bearing names like Greenland, Hobbs, Tuckney, Prank's Stone and Farview. For days they rode, always alert, and by keeping a steady pace, they reached the area south of Cavell Village. Several times they had passed bands of armed men, but none had offered them challenge, and they arrived without incident.

Rounding a bend in the road, they crossed a small bridge that took them over a swift-running stream. James looked down and observed, 'This is deep.'

Owyn said, 'Deeper than it looks. More than one idiot's been drowned trying to swim across. It's a feeder to the River Rom, coming down from the mountains over there.' He pointed to the west, where bluffs rose. 'Let me show you something,' he added as he turned his horse off the road.

They followed an old dirt roadway, grown over by grass in several places, obviously unused for a long time. Gorath said, 'I see fresh tracks. Someone has ridden here lately.'

Owyn said, 'Undoubtedly. I'll show you why when we round this bend.'

They rode around a sharp turn, where a bluff rose up to a cliff-top overhead, and halted. Before them an impressive-looking waterfall thundered down from the cliffs above, exactly three hundred feet above. On both sides the gorge rose steeply, and was covered with thick forests.

'Cavell Run,' said Owyn.

'What's that?' asked James.

'It's the name of the stream. It's also what we call the

tunnels under the old keep.' He pointed to the top of the cliffs and by squinting James could make out the grey edifice that rested atop the cliffs.

'How did you know about this?'

Owyn turned his horse back and said, 'When I was a boy, we came here several times. I used to play with my cousin Ugyne in the run. They're a huge set of tunnels and caves under the keep. Used for storage in ancient times, but mostly abandoned now.' He pointed backwards as they left sight of the waterfall. 'There's even a bolt-hole behind the waterfall if you know where to look. Ugyne and I found it from the inside of the run when I was nine and she was eight. We stripped off and went swimming. We almost froze to death; the water is all snowmelt running down the ridges from the mountains above. Ugyne got a pretty heavy whipping from her father, too. My uncle has never curbed his temper as long as I've known him.

'But it still didn't stop Ugyne and me from playing up there.'

James asked, 'How many know about the run?'

'Most of the locals know there are tunnels under the old keep. A few might even suspect there's a bolt-hole under the waterfall. But I doubt anyone outside the family, the old guard commander, and maybe one or two of the older servants, has any idea where it is. It's pretty well hidden.'

They continued on toward Cavell Village, arriving at mid-afternoon. As they turned off the road and moved to within sight of the place, James said, 'For a village it's rather prosperous.'

Owyn laughed. 'I guess. It was a village for a couple of hundred years, but became a busy farming centre about fifty years ago. Since the fire in the keep forced my uncle to move into the village about three years ago, all business is conducted down here. I think he and his household account for a third of the houses here in the village.'

'Fire?' asked Jimmy as they reached the outer buildings. 'What was that?'

'No one knows,' said Owyn. 'The story is my uncle was having some work done in one of the lower chambers and a fire broke out, working its way up through the building, gutting it and making it unsafe to live in. There had already been a collapse in the lower tunnels, where my uncle was expanding his wine cellar. My cousin Neville died in that collapse. He was a few years older than Ugyne and me. He was an odd boy; it always seemed to me his father didn't care much for him. Ugyne was always Uncle Corvallis's favourite.' He was lost in memory for a moment, then returned to the present. 'Anyway, that basement was just sealed off, with my cousin's unclaimed body still under tons of rock.

'The fire started not far from there, and the maid who is blamed for starting it died in the flames, so no one is quite sure how it began. It burned up from below, weakening timbers and causing floors and walls to collapse. Uncle's been telling everyone he was going to repair everything and move back in some day, but so far we've seen little proof of it.'

They rode down the main street of the village, a broad thoroughfare that ended in a large square, dominated by a fountain and three other streets which ran off at odd angles to the one on which they rode. 'That house over there,' said Owyn, turning his horse so they could ride around the fountain. The afternoon market was underway and the buyers and sellers ignored the three riders for the most part, though one or two gave Gorath a second glance.

They reached the front of the Baron's house and a stable-boy ran over and said, 'Master Owyn! It's been years.'

Owyn smiled. 'Hello, Tad. You're caring for horses now?'

The boy, no more than twelve or thirteen years old, nodded. 'Yes, sir. Now that we have no proper stable the Baron's keeping his guests' mounts over at the inn.' He pointed to an inn directly opposite the Baron's house. It was dominated by a sign of a wood-duck's head. 'I'll arrange rooms for you.'

Owyn smiled. 'You're telling me my uncle won't be happy to see me and offer me a room?'

The boy nodded. 'He's not really happy to see anyone, these days, Master Owyn. If you were here alone, he might offer, but with your friends . . . ?' He smiled apologetically and said no more.

Owyn sent him off with the horses and instructions to get them one large room for the night.

They mounted steps to the large house. James glanced around and said, 'This house dwarfs the rest in the village.'

Owyn smiled at the understatement. The rest of the village ranged from simple huts of wattle and daub with thatch to some two-storey wooden houses with small gardens. The inns were the only buildings that matched the Baron's residence.

'It used to be an inn, but fell on hard times. My uncle bought it and converted it to his own use. There is a stable in the rear, but it's occupied by his company of personal guards.' Lowering his voice, Owyn said, 'Like many minor nobles, my uncle has more rank than money. The rents are modest, the taxes to the Duke of Cheam considerable, and my uncle has never been what you would call an enterprising man.'

They knocked upon the door. The door opened a crack. A serving woman of middle years peeked through and when she saw Gorath in his armour standing before her, her eyes widened and her complexion turned pale. 'Hello, Miri,' said Owyn, coming into her field of vision. 'It's all right. They're with me.'

The woman said, 'Master Owyn,' and swung the door wide.

'Could you please tell Uncle Corvallis we're here?'

The woman nodded and hurried off. A few minutes later a

tall man, affecting a velvet coat and lace-front shirt, with far too many rings, arrived and said coolly, 'Nephew, we had no word of your arrival.' He cast a disapproving eye upon James and Gorath.

'That's all right, uncle. We intrude. We've already made arrangements to stay at the inn across the square. May I present to you Seigneur James, squire to Prince Arutha, and our companion, Gorath. Gentlemen, my uncle, Baron Corvallis of Cavell.'

At the mention of a relationship to the Prince of Krondor, Baron Corvallis's attitude softened slightly. He nodded at James and said, 'Seigneur.' Looking at Gorath as if he didn't know what to make of him, he said, 'Elven sir, welcome.' He made a sweeping gesture and said, 'If you will join me in my parlour, I'll send for some wine.' He signalled to the serving woman and said, 'Miri, a bottle of wine and four goblets.'

They followed the Baron into a hallway through what had been the old common room of the inn, now divided into several different rooms. The rear stairway to the upper rooms was visible at the end of the entrance hall, and James absently wondered if the old bar was still intact. Apparently he would never know, as they turned into a corner room with two large windows, overlooking the village square. The Baron indicated three chairs and took a fourth for himself. 'What brings you to Cavell Village, Seigneur?'

'The Prince's business,' said James. 'There was some trouble down in Romney, and, as an outgrowth of that, we're investigating rumours of Nighthawks returning to the Kingdom.'

At mention of Nighthawks, the Baron almost levitated out of his chair. 'Rumours!' he shouted. 'They are not rumours. There is wicked slaughter being done here in the north and I have sent reports to my lord the Duke of Cheam. They have tried to kill me three times!'

James attempted to look concerned. 'It was those very things that brought me here. The Prince is adamant, as is his brother the King – ' Lyam probably had no idea what was happening, but James had long ago learned that dropping the King's name from time to time was a very powerful thing to do ' – can't countenance the idea of unprovoked assaults upon their nobles.'

At mention of the King, the Baron seemed almost reassured. 'Good, it's about time.'

James said, 'Why don't you tell us of your situation.'

His face flushed with emotion, the Baron spoke quickly and with anger. 'Three years ago a maid died in a fire that started near the abandoned wine cellar. At the time, I thought it was merely a tragic accident, but now I'm convinced it was but the first attempt on my life.

'A year ago, while out hunting, a band of riders, all clad in black, appeared on the ridgeline and rode at us with weapons at the ready. Only a fox flushed by my hounds saved me, as the animal bolted across a field before the attackers, and the pursuing hounds caused their horses to falter. Lost my best hound that day.'

He motioned to Miri, who had appeared at the door, to serve his guests. 'Then last month, I was shot at by men from behind cover. The arrow cut my tunic, here.' He pointed to his shoulder. 'A hand's span lower and I'd be a dead man.'

James glanced at Owyn who nodded slightly, indicating the Baron wasn't exaggerating.

Baron Corvallis continued. 'I dare not leave my own house, save perhaps to visit the inn with personal guards on all sides. My daughter disobeys me and runs like a common child across the fields and consorts with all manner of questionable riff-raff. She should be meeting respectable suitors at her age, but instead she walks through the fields with . . . a despicable creature who woos her with sweet lies.'

Owyn tried to look serious, but was obviously amused by something. He said, 'Who is this foul being, uncle?'

'A man of commerce! Ugyne should be accepting court

from the sons of barons, earls, even dukes, but not a common merchant. My solicitor Myron loves her, and while lowborn, has some ties to nobility. I would suffer him ask for her hand if she would settle down, but she's filled with fanciful notions of romance and adventure, irritating enough traits in a son, but utterly unacceptable in a daughter.'

'Does this agent of chaos have a name, uncle?' asked Owyn.

Nearly spitting, Corvallis said, 'Navon du Sandau! I know he is a criminal. He wears clothing of costly weave and rides the finest black horse I have seen, yet he speaks little of his commercial enterprises. He claims to be a factor for several rich families and nobles, as well as an agent for trading concerns in the south and west. Yet I have never seen him on an errand of business; rather he is mysteriously absent or hanging around, wooing my daughter.'

Owyn sipped at his wine, then asked, 'Where is Ugyne, uncle?'

'Probably out near the road, wandering the fields, waiting for snow to fall or Navon to arrive.'

James took another drink of the somewhat indifferent wine and said, 'We've imposed upon your hospitality long enough.' He stood and said, 'We'll investigate this as quickly as we can and see what can be done to end these threats on the peace of your village.'

'Thank you, Seigneur,' said the Baron. He said, 'Owyn, give my regards to your father and mother when next you see them.' He nodded at Gorath as the moredhel walked past. Unsure of what to say, he merely nodded again.

At the door, he said, 'Owyn, if you're in the village next Sixthday, do me the pleasure of dining with us. Bring your friends.'

The door closed and James laughed. 'That gives us five days to find what we're looking for and leave before he's forced to make good on his offer.'

Owyn said, 'My uncle is a difficult man at the best of times, but he is genuinely frightened.'

'Even I, who know not your race that well, could tell that,' said Gorath. 'Yet one thought bothers me.'

'What?' asked James. 'Only one?'

'Among many,' said Gorath. 'If the Nighthawks had truly wanted him dead, he would be dead. The dogs interrupting the attack on horse, perhaps. But a near miss by an archer seems improbable.'

'Having faced the Nighthawks several times, I'd agree,' said James. They entered the Duck's Head Inn.

The common room was relatively uncrowded, it still being afternoon. The innkeeper crossed from behind the bar and said, 'You're the gentlemen in to see the Baron?'

'Yes,' said James.

'I'm Peter the Grey,' he said with a slight bow, 'and I have the privilege of owning this establishment. Your rooms are ready any time you are, and we have a full board and a choice of wines and ale.'

'Ale,' said Gorath. 'I have little affection for wine.'

James laughed. 'Given the Baron's choice in wine, I don't blame you.'

Owyn nodded. 'You can't imagine what it would have been had you not been a member of the Prince's court.'

Peter the Grey's eyebrows shot up. 'A member of the Prince's court? Well, then, I best ensure we only serve the finest. A *member*, gentlemen!'

As Peter hurried away, James called after, 'And food, please.'

They sat and Owyn said, 'Sorry you had to endure the ramblings of my uncle. Compared to the troubles we're investigating, his woes must be pathetic by comparison.'

James was thoughtful. 'Perhaps, but there may be a connection here. I'm not quite sure what it is, but why would the Nighthawks harass your uncle, yet not kill him?'

'To keep him frightened,' suggested Gorath.

Just then Peter the Grey arrived with the ale and placed frosty mugs before each of them. James sipped and nodded with appreciation. 'Wonderful.'

'Ale from the Grey Towns, sir, and we keep it cold.'

'You ship ice down here?'

'No,' said Peter. 'There are deep caves not too far from here where I leave my barrels. I sell it too quickly for it to warm up before the barrel's empty.'

James smiled. 'Situated as you are directly across the square from the Baron's home, you must see him a lot.'

Peter shook his head. 'Hardly at all, truth to tell. The Baron only leaves his home rarely, and then always with armed guards.' He picked up his tray and said, 'I'll bring some food straight away, sir.'

James said, 'Something is eating at my mind, but I can't quite pin it down.'

'Something to do with my uncle?'

'Yes,' said James, 'but Gorath has pointed out the one thing in this that makes no sense: why go to the trouble of frightening the Baron, but not kill . . .' Suddenly James's eyes widened. 'Peter!' he called.

The patron of the inn returned in a hurry. 'Sir?'

'What was it you just said about the Baron, about you not seeing him.'

'I just said the Baron leaves his home only rarely, and then with armed guards.'

'When did this start?'

'Right after the Nighthawks started hunting him, I guess.'

'You know about the Nighthawks?' asked James.

'Well, we know what people say.'

'And what would that be?'

'That the Guild of Assassins has set up shop around here and they've marked the Baron for some sort of punishment.'

James said, 'Thank you, Peter. Sorry to have disturbed you.'

Owyn said, 'Why did you want him to come back?'

'To help me think this through,' said James. 'Look, the Nighthawks aren't trying to kill the Baron. They're trying to make the Baron stay in his house.'

'Why?' asked Gorath.

James said, 'To stop him from rebuilding the keep.'

Owyn said, 'What in the world good would that do? It's an old fortification, and if there's an army heading this way, it's not going to cause them much trouble.'

James said, 'I don't think anyone cares about the keep. I think they care about what's under it.'

Owyn's eyes widened. 'The run?'

'You said there was a secret passage into caverns that run under the mountains, and the old keep's armoury and storage are down there. You could hide an army under there, I bet.'

'Or a nest of Nighthawks,' added Gorath.

Owyn said, 'But how would they know?'

'The run isn't a family secret is it?'

'No, a few others know of its existence, but finding the entrance from outside would be nearly impossible.'

'Owyn!' a female voice cried happily from across the inn.

They turned to see a tall, leggy young woman in a simple dress hurrying across the room. She nearly knocked Owyn back into his chair as he tried to stand up while she threw her arms around him.

'Uh, Ugyne!' said Owyn, grinning and blushing at the same time as she hugged him.

The girl was pretty in a sunburned, wild fashion. Her hair was windblown and unkempt and she looked as if she had been sitting on the ground, as her dress was streaked with dried mud in the back.

She stopped hugging him long enough to deliver an enthusiastic kiss on the lips, then she stood back, holding him at arm's length as she studied her cousin. 'You've grown into a fair-looking man, given what a pathetic little boy you were,' she said with a laugh.

Owyn blushed deeply and laughed. 'You haven't changed, I see.'

She pushed him back into the chair then sat down imperiously on his lap. 'Of course I have. I was a little girl the last time you saw me; now I'm a grown woman.'

James grinned. This grown woman appeared to be eighteen at the outside, and while she was striking in her vivaciousness, she was still a little gangly and moved with a studied purpose, as if to mask her uncertainty.

Owyn said, 'Ugyne, these are my friends, James and Gorath.'

She nodded and smiled as she said, 'Hello.' Of Owyn, she asked, 'Have you seen Daddy yet? I assume you did. Tad was the one who told me you were here.'

'We did, and if we're here on Sixthday, we'll be dining with you.'

'Oh, please do stay. Supper alone with Father is such a bore.'

James said, 'We may be gone, Ugyne. We have pressing business.'

'What sort of business?' she asked with a pout. She looked at Owyn. 'My favourite cousin comes to town after too many years and wants to bolt the next day?'

Owyn said, 'No, but we're on . . . business for the Crown.'

'Oh?' she said with raised eyebrows. 'Really?'

James nodded. 'Really.'

'Well, then,' she said, 'I'll have to insist that either Father have you over earlier or you stay, but you'll not leave town until we've had a chance to visit.'

'What have you been doing with yourself?' asked Owyn. 'Your father seems very concerned about how you spend your time.'

She turned up her nose at the mention of her father's opinion and said, 'Father wants me to sit around all day in that dark house, waiting for some noble to ride up and take my hand in marriage, and is terrified I'll run off with someone.'

'Anyone in particular?' asked Owyn.

She reached over and took his mug of ale and took a delicate sip from it, as if it was the most brazen act imaginable. 'There's Myron, Daddy's solicitor here in the village. He's a widower with a lovely little girl I adore, but he's so . . .'

'Dull?' supplied James.

'No, predictable. He's a nice man, but I want something more.'

'Anyone else?' asked Owyn.

'Why? Did I say there was anyone else?' she asked with a glimmer in her eyes and a smile on her lips.

'No,' said Owyn, 'but your father did.'

'Navon du Sandau,' said Ugyne. 'He makes Father furious.'

'Why?' asked James.

'He's a man of trade, not nobility, and even Myron, my father's solicitor, is related to nobility: he's the nephew of the late Earl of Silden, on his mother's side.'

'Are you in love with Navon?' asked Owyn.

She shook her head and wrinkled her nose. 'Not really. He's interesting, if a little . . . strange.'

'Strange?' asked Owyn. 'How?'

'I find him staring at me in odd ways, when he thinks I'm not looking.'

Owyn laughed and tickled her. 'That's because you are odd looking.'

She playfully slapped his hands away. 'But he's interesting. He's very attractive, and intelligent, and he says he's been everywhere. And he has a great deal of wealth, which is the only reason Father hasn't ordered him whipped out of town by the guards. If I can't marry nobility, Father will settle for wealth.'

'Are you going to marry this Navon?'

'Probably not,' she said, jumping out of Owyn's lap. 'He's too ardent and . . . dangerous.'

'Dangerous?' said Gorath, speaking for the first time to the girl. 'I know little of your customs, but isn't that an odd term to describe a suitor?'

She shrugged and replied, 'I don't know. He's fascinating, if a little odd at times, and he's taught me a few things.'

'Oh?' asked Owyn, his voice registering both curiosity and disapproval.

She punched him in the shoulder. 'Not that, you evil boy! He's taught me about things like poetry, music, and he's taught me to play chess.'

'Chess?' asked Owyn, casting a glance at James.

'Yes,' she said. 'He's the finest chess player in Kenting Rush, probably in the entire area. He travels to Malac's Cross regularly to play against the best in the Kingdom at the Queen's Row Tavern and has played against nobles in Krondor and Great Kesh!' Her description indicated some pride in the claims.

'Well,' said James. 'Perhaps we can meet him some time.'

'Come to supper on Sixthday and you can,' she said. 'He's coming to see me by the end of the week!'

With a laugh and a half-twirl that set her skirt swirling around her knees, she turned and half-skipped, half-walked to the door. Looking over her shoulder, she smiled at Owyn and left.

Gorath said, 'The women of your people are . . . interesting.'

James laughed. 'She's young. She's working a little too hard at being vivacious.' He shook his head in appreciation. 'But give her a couple of years and she won't have to work. She's quite the charmer.'

Owyn sighed as he leaned back in his chair. 'She's the only member of my family I ever really cared for around here.'

Peter the Grey arrived with their food and as he set the

table, Owyn said, 'I never knew my cousin Neville – he died when I was young – and I had only seen him once before that.'

Peter interrupted, 'Baron Corvallis's Neville? You said you were in to see him, young sir, but nothing about being his nephew.'

'Sorry,' said Owyn. 'I wasn't trying to hide the fact.'

'You're young Owyn,' he said. 'You don't remember me, do you?'

Owyn said, 'Sorry, but I don't.'

'I was one of the cooks up in the keep, before that tragic day when young Neville died. You were only six or seven back then, and I only saw you once or twice when you visited. I bought this inn not long after, and you never stopped in before today. The old Baron, well, it changed him. He was a different man after that, but it killed his wife.'

'I don't remember much about it,' admitted Owyn.

Peter needed little prompting to gossip and said, 'Well, the story goes that there was some difficulty between the Baron and the master builder he hired to work on the lower caves and tunnels as he expanded his wine cellar. The odd thing was he was also named du Sandau, like Navon.'

James and Owyn exchanged glances.

Peter went on. 'Well, this Sandau was the finest stone-mason in the region, but he was also a drunk and a wom-anizer; rumour is he had his way with many of the ladies of the court down in Rillanon before coming north.

'He worked on several portions of the old run, under the keep, and usually the Baron was happy with the work. But this wine cellar, for some reason, had problems. They argued and the Baron was always in a foul temper.

'Then came that black day.'

'The day Neville died?' asked Owyn.

'Yes, it was the same accident that killed Sandau. The ceiling collapsed. No one knew why. All the men in the area struggled for days to remove the rubble, but it was to no avail; Neville and the workers in the room died.'

'What was the boy doing in the room?' asked Gorath.

'No one knows. He liked to watch the masons, and his father didn't object.' Peter shrugged. 'But the Baron's never been quite the same since then. And the loss of the boy killed the Baroness, I will avow. She mourned for months, then got sick, and even the healing priests from the temples couldn't keep her alive. She died a little more than a year after. Before the boy died, she was a woman of unusual steel. Ugyne's like her; it's what kept the girl sane, I think, losing a brother and mother within a year.' Peter shook his head in sympathy as he recalled the girl's pain. 'She's managed to turn into quite a special person, by my lights.'

James nodded as Owyn said, 'She is, no argument.'

Peter left and James said, 'This family of yours has had its share of tragedy.'

Owyn said, 'I know. But Ugyne seems to have found some happiness.'

'Even if it's only tormenting her father,' said James, and even Gorath laughed at that.

'Well, then,' asked Owyn, 'what are we to do?'

'I think we have dinner with your uncle on Sixthday and I think we see if someone here wants to play chess.'

Owyn nodded and sat back, content to rest a few days before the next conflict.

TEN

Nighthawks

Water thundered down the mountainside.

James, Gorath and Owyn sat on their horses near the base of the falls. With a few days to fill in between their discussion with Ugyne and their coming supper with her father on Sixthday, James had decided to scout around. He had made sure the talkative Peter the Grey knew they were heading down the road on business, but as soon as they had cleared the precinct of Cavell Village, they had turned off the road to investigate Cavell Run.

The spray struck James as the wind shifted. 'You used to play here?' he asked Owyn.

'No, not really.' He pointed up the side of the mountain. 'We used to play up there, in a pool, near the spot the bolt-hole exits the hillside.'

Gorath said, 'My people's children are not allowed to play unsupervised.' With a note of contempt, he added, 'But then you humans breed like fieldmice; if a child dies, you just have another.'

James threw him a black look. 'It's not quite that simple.'

Gorath asked, 'So why are we here?'

James asked, 'If you wanted to use the old run as a base of operations, would you want the Baron and his family up there?'

Owyn's eyes widened. 'You think the Nighthawks started the fire?'

James shrugged. 'I don't know. But it's pretty convenient, and by harassing him, they keep the Baron from starting his rebuilding.'

They rode along the banks of the river toward the cliffs, and Gorath said, 'I have fought these Nighthawks at your side, and you have mentioned them before, but I still do not understand their part in all this.'

James said, 'It isn't difficult; they're a brotherhood of assassins who work for whoever pays their way. Mercenaries. I faced one on the roofs of Krondor when I was a boy and have faced them many times since then.

'They were pawns of Murmandamus for a while and served with his Black Slayers.'

Gorath almost spat. 'The Black Slayers were an obscenity! Men of no honour who gave over life and spirit to Murmandamus for promises of eternal power and glory! It is said by our lore keepers that those who did so will never join the Mothers and Fathers in the Life After.'

James turned his horse to follow around a small knoll, and said, 'I must admit, I know little of you or your elven kin, Gorath, though I've fought the moredhel and spent time with the glamredhel and elves.'

Gorath said, 'We dislike one another enough that we don't like to talk about one another, it's true, so I have no doubt you heard little good of us from the eledhel. The glamredhel are the mad ones, those without purpose and without magic. They lived by their wits and held strong in the Edder Woods in the Northlands until they were hunted down and destroyed.'

James shook his head. 'Destroyed? They've gone to Elvandar and now reside there.'

Gorath reined in his horse. 'Delekhan!'

'What?' asked James, turning to look at the dark elf.

'He let it be known that he had destroyed Earnon and his tribe in the Edder.'

'Well, Old King Redtree is alive and well, living up in Elvandar. Last I heard they were involved in some sort of discussion as to who was in charge.'

Gorath tilted his head, as if listening to something. 'In charge? I do not understand.'

'I don't pretend I do, either,' said James as they followed another bend in the road, and began approaching the waterfall. 'Duke Martin is a regular visitor to Elvandar and sends reports to Krondor. As I understand it, Redtree and his people are trying to decide if they're going to be part of Aglaranna's people, or separate, but living among them. Something like that.'

'It's passing strange,' said Gorath. 'I would assume Aglaranna would enslave them had they come begging for refuge.'

James laughed.

'You find that funny?'

'I've met old Redtree and he doesn't exactly strike me as the type to beg or to accept slavery without killing a couple of hundred people first.'

Gorath nodded. 'He is a warrior of great skill and power.'

They could again feel the spray off the waterfall and James asked, 'Owyn, where is the entrance?'

Owyn said, 'We'll have to tie the horses and walk from here.'

They did so, and as they reached a place beside the waterfall, where the spray was heavy enough to soak them in minutes, James said, 'How many people knew of this entrance?'

'A few, in my family, and among the staff. Ugyne and I, along with Neville, used to play there. We got beaten when we were caught, and I don't think the Baron ever found out that we knew the entire route from the keep to the bolt-hole.' He pointed to a rock a few feet above his head. 'This is why no one in the village ever found their way into the keep. I need a leg up.'

James made a cup with his hand and gave Owyn a boost, and the young magician pulled himself to the ledge. He said, 'Hand me my staff.' They did, and he said, 'Now, stand back.'

They stood away, and Owyn used his staff to move a rock.

A rumbling caused James to move even farther back. A large rock face moved aside. Owyn jumped down with an 'oof' and stood up. 'Getting out's easy. There's a lever just inside. Getting in is impossible if you don't know the trick.'

James moved just inside the entrance and said, 'Someone found the trick. Look.'

Dust had coated the entire length of the tunnel, but the middle of the tunnel showed clearly that many feet had trodden the floor recently. Gorath said, 'As we move along this tunnel, we will soon lose the masking noise of the waterfall. Tread softly.'

James said, 'We need a torch.'

Owyn said, 'No, we don't. I'll make us some light.'

Owyn closed his eyes, then held out his hand. A sphere of soft light surrounded him, less than would have come from a torch, but enough for them to see by. 'That's handy,' said James.

Owyn shrugged. 'Until recently I didn't know if I'd ever use it for anything more significant than finding my way to the jakes in the middle of the night.'

James grinned. 'Let's go.'

He pulled out his sword as did Gorath, and without a word they set off down the tunnel.

A soft tread of boot leather on stone was all Gorath needed to warn them. He held up his hand and listened, his more-than-human hearing announcing the approach of someone. He turned and held up two fingers.

James nodded and motioned for Owyn to move back down the tunnel, taking his faint light with him, while he and Gorath waited in the gloom for whoever came toward them. A moment later a light could be seen down the hall, approaching rapidly. Voices echoed off the rock.

'I don't like it,' said one.

'You don't have to like it. You only have to follow orders.'

'There used to be a lot more of us, if you remember.'

'I remember, but the fewer of us, the more gold – '

The two men turned the corner and Gorath and James leaped upon them. Catching them unexpectedly, James and Gorath had them down before they knew they were under attack.

But surprise didn't mean surrender, and the two assassins fought like cornered animals, forcing Owyn to run forward with his staff and lay one low with a crushing blow to the head.

The other died upon his own knife, as James fell heavily atop the man.

James slowly rose, saying 'Damn. I wanted a prisoner.'

Gorath said, 'We are in their nest. It would be wise for us to leave now that we know where they are and return with soldiers.'

'Wise, perhaps, but my experience with these birds is they will have flown by the time we return. They are never abundant in number, and quite a few have died recently. I doubt there are more than a half-dozen left between here and the Teeth of the World.' James pointed a finger down the hall from where the two had come. 'But if we identify or trap their leader, we may finally be done with this bunch.

'I thought them dead and buried ten years ago, but obviously I was wrong. At the least one or two of them fled to start this murderous brotherhood again. Only fanatics kill themselves like that. I must find out if these are but hired blades working for whoever pays the most, or if they are willing allies of your Delekhan.'

'What difference does it make when it's Kingdom throats being cut?' asked Gorath.

'Men who work for gold are one thing. Men pledged to dark causes are another. If these are men working for gold, we can deal with them at leisure, for they will know little beyond where to pick up their gold and whom to kill. But if they are involved in these dark plots, perhaps we will learn something – ' he pointed down the hall ' – down there.'

Gorath and Owyn exchanged glances, and Owyn said, 'Well, I'd get bored out there waiting for you to come back.' He held up his glowing ring. 'Besides, I have the light.'

Gorath gave a grunt that might have passed for a chuckle.

For nearly half an hour they walked through a long tunnel, then Owyn said, 'There's a storage room ahead, if I remember.'

They found a large wooden door, still intact and well oiled, behind which was a barracks. A score of beds were lined up, ten against each wall, and racks of weapons occupied the far end of the room. Most of the beds hadn't been slept in, but four showed recent occupation. Owyn pointed and whispered, 'Those two we killed may have friends close by.'

'Or they could already have left,' said James.

They moved to the racks and saw that the weapons were polished and ready. A variety of lethal-looking blades were stored in orderly fashion, as well as daggers, throwing knives, darts and strangling cords. A shelf full of jars was attached to the wall above the rack. 'Poisons, I'm willing to bet,' said James. He looked at Owyn. 'How much further do these tunnels go?'

'Miles, if you mean all the levels. This is the lowest gallery, and there are three between this one and the basement of the old keep. Though I don't think we could get there because of the caved-in wine cellar.' He pointed to a door at the opposite end of the room. 'Through there is another room like this one, and then stairs up.'

James went to the door and listened. Hearing nothing, he opened it and found another barracks, with twenty well-made empty beds. 'No one has been here for a while,' he observed.

'Not quite true,' said Gorath, pointing. 'One pair of footprints. Heading that way.' He indicated the far end of the room, where stone stairs rose up the wall to a hole in the ceiling. Next to the stairs was a bed left unmade, apart from the others. A huge wardrobe had been placed next to the bed, incongruous in its setting. It was made of highly polished wood with gilt trim, and when James opened it, clothing of expensive weave and boots of fine leather could be seen.

'I'm willing to bet the leader of this band of cutthroats is the dandy who uses this bed.' He looked around. 'See if there's anything here that might identify this fashion pate. I'm going to check the next floor.'

James hurried up the steps and discovered a large wooden door barred the way. It was attached to the stones by heavy hinges and a hasp with a lock. Locks had rarely proved a problem to the former thief, but this one was of ingenious design and James had fallen out of the habit of travelling with lockpicks. 'Owyn, what's up here?'

Owyn paused, as if searching his memory, then said, 'It's another storage room, smaller, but similar to this one, and then there's a long tunnel leading back into the mountain.'

James came down the stairs. 'Either our quarry is hiding something up there from his own men, or he's fearful of someone stumbling into this lair from above.'

'I doubt the second case,' said Owyn. 'Someone would have to get into the old keep, know how to activate the door from the armoury to the first tunnel, and besides, most of the upper passages were buried in the collapse of the old wine cellar.'

'Then he's keeping something under lock for his own reasons.'

'Perhaps gold,' suggested Gorath. 'Assassins would have to be paid.'

James said, 'There is that.' He came down the stairs. 'Find anything?'

'Just this,' said Owyn. He held out a book.

James took it and read the title on the first page. '*The Abbot's Journal*,' he read aloud. He flipped a few pages and said, 'It's a collection of stories about your uncle's family, it seems.' He handed it back to Owyn. 'How did it get here?'

Owyn said, 'I have no idea. It may have been lost when my uncle evacuated the keep after the fire, and someone combing the rubble above might have found it.'

'Bring it along,' said James. 'I think I'll do some reading before bed tonight.'

James led them back the way they had come.

James moved the beds and Gorath asked, 'Is this some human custom of which I'm not aware?'

James grinned. 'Unless there were no other Nighthawks around, someone is going to find it odd that two of their lads went missing. My best guess is Nighthawks don't usually go absent without permission. So it's not unlikely that whoever discovers they are gone might decide to come see if we three had something to do with it.'

Once he had the beds crowded against the door, he said, 'If they act as usual, one or more of them will come through that window while the rest come in that door. They'll come fast, through the outer door and up the stairs before Peter the Grey can get out of bed to find out what's causing all the noise. If they work as planned, by the time old Peter gets through the kitchen and up these stairs he'll find three bodies here and an open window.'

Owyn said, 'If they come.'

James grinned. 'Oh, they'll come. We're the only new-comers in the area who've been hanging around, visiting the Baron, asking questions. I just don't know if they'll come tonight or tomorrow night.' James turned the lamp down low, enough so he could read, sat down next to the lamp on the floor, and opened the book he had had Owyn carry back from Cavell Run.

Owyn produced a second book and said, 'I might as well put this time to good use, as well. I've neglected this too long.'

'What is it?' asked Gorath.

'My book of magic.'

'You wrote a book?' asked the dark elf.

'No, it's a book each student keeps, recording thoughts, discoveries, and notations of things observed or learned.' He produced a quill and a tiny vial of ink. 'When Nago almost hit me with that spell he threw, I sensed something, and, well, it's hard to explain, but I'm puzzling out how he did it. I think with some more study I can do it.'

James looked up. 'What does it do?'

'If I'm right, it should immobilize the person struck, maybe more.'

'More?' asked James, now very interested.

'I think it might eventually kill the victim.'

Gorath, said, 'If it immobilizes, what does it matter? You just pull out your dagger and walk up and cut his throat.'

Owyn said, 'I guess. When I was at Stardock, the teachers didn't delve too deeply into violent applications.'

James yawned. 'Which is wise. It wouldn't do to have a bunch of you youngsters wandering around that island tossing off fireballs and blasts of lightning at each other. The tavern brawls would be pretty impressive in the carnage they left behind.'

Owyn laughed. 'Maybe you're right. Some of the students were twice my age. I think magic takes a long time to master.'

James said, 'If one ever does master it.'

'I heard Pug was a true master,' said Owyn.

James yawned again. 'I've seen him do some pretty impressive things,' he admitted through his yawn. 'Mercy, but this waiting is trying on the nerves.'

'Then get some sleep,' said Gorath. 'I'll watch.'

Owyn asked, 'Do you know Pug well?'

'We met a few times,' said James. 'Why? Didn't you meet him at Stardock?'

'No, I saw him from time to time, with his family, but he spends most of his time in his tower or off away from Stardock. Most of the teaching is done by others. I met him only that one time in Krondor, briefly, when his daughter was trying to read Gorath's mind.'

'I've never met the girl, though I hear she's a nice kid,' said James, as he thumbed through the book. 'Her brother Willie's a good lad. He's training to be an officer in Arutha's guard.'

'Hmmm,' said Owyn, and James glanced over to see the young magician lost in his notes.

James looked through the book in his lap again for nearly a half hour. 'This is the most improbable collection of accounts and . . . outright fabrications I've ever encountered.'

Owyn looked up. 'What do you mean?'

'There's lists of births and deaths, as if someone sat down one day and told this Abbot Cafrel the Cavell family history in one sitting, then suddenly we're talking about missing treasure, swords of incredible magic power, and curses.'

'Sounds interesting,' said Gorath, who was trying to be polite.

James laughed. 'I agree,' he said, putting aside the book. 'You watch and I'll sleep. Wake me in two hours.'

James curled up and Owyn studied, and Gorath watched the window, his hand resting on his sword.

They came the next night. James had again been reading the Cavell family history and Owyn was meditating on the bed, his eyes closed as he was developing a method of casting the spell Nago had used on him. Gorath lay sleeping on the floor, having elected to sit the later watch.

One moment James was reading, and the next he was moving, his sword coming out of his scabbard. Owyn was shot forward by two heavy bodies hitting the other side of the door as the window shutters exploded inward. An assassin had tied a rope to the roof beam and swung out, so he could crash feet first through the wooden shutters into the room.

He caught James full in the chest and the squire flew

backwards into Gorath. Owyn came up on his knees, then fell back out of the way of a sword blow, while behind him someone was trying to force the door open.

Owyn had been halfway through constructing the spell in his mind when suddenly letters of fire seemed to burn in his mind's eye. He raised his hand and pointed it at the assassin who was again raising his sword. An evil purple-grey sphere, black veins of energy dancing across its surface, leaped from his hand, striking the assassin in the face. The man froze as if suddenly transformed into purple stone, blue sparkles of energy dancing across the surface of his body. A faint moan of pain escaped his lips.

James was up and ran to the window, thrusting his sword through it as another man tried to swing in. The second Nighthawk was impaled on the blade and fell into the stable yard below, striking the stones with a sickening wet thud.

Gorath regained his footing and threw his weight against the door. He shouted, 'Do we try to hold the door?'

James said, 'When I yell, jump back and pull that last bed with you.'

Owyn was staring at the entranced assassin in wide eyed wonder. 'It worked!' he whispered.

James struck the ensorcelled man as hard as he could across the back of the head with the flat of his sword and he crumpled to the ground, the energy around him vanishing. 'Can you do it again?'

'I don't know!'

'Then get out of the way! Gorath, now!'

Gorath did as he was told, and Owyn grabbed the bed and pulled it away as well. The other two beds began to slide away from the door.

'If I know my Nighthawks,' said James. 'I suggest you duck . . . now!'

Both men did so as James fell to the floor. The door burst open and two crossbow bolts flew into the room and vanished out the window. James instantly jumped atop the bed Gorath and Owyn had just moved. He bounced off the bed and crashed into the two men closest to the door, sending them through the railing of the stairs to the floor below. He slid over the edge of the landing, barely avoiding a fall by grabbing a part of a shattered post. His sword went clattering to the floor below, as an astonished and shocked Peter the Grey entered the room from behind the bar. 'What?'

James looked up from where he hung to see a Nighthawk standing over him, sword raised high. The assassin's eyes went round as Gorath ran him through with his sword. The last Nighthawk tumbled over James to the floor below, landing at Peter's feet.

'Oh, my word!' said the innkeeper. 'My word!'

James hung by one hand and said, 'If it wouldn't be too much trouble . . .'

Gorath's powerful hand seized him by the wrist and hauled him up to the landing. James said, 'Thank you,' and hurried down the stairs, rubbing his sore shoulder. 'I'm getting too old for that sort of thing,' he observed.

'What is going on?' asked Peter.

James knelt next to the last assassin and began searching the body. 'These men tried to kill us,' he answered calmly. 'We didn't let them.'

'Well . . .' said the innkeeper. 'Well . . . I . . .' After a moment, he said, 'Well,' one more time.

James said, 'Get somebody in here to clean up the mess, Peter. Else your customers may be put off their meals.'

The innkeeper turned and hurried off to do as he was bid. Instructions like that he understood. To Owyn, James said, 'You'd better go get your uncle and explain to him that we've just removed most of the Nighthawks who were stalking him.'

Owyn said, 'I think he might not even object too much to being awakened in the middle of the night for that bit of news.'

After Owyn left, Gorath said, 'I noticed you said, "most of the Nighthawks who were stalking him".'

James stood up, after having found nothing useful on the bodies. 'We still have one Nighthawk to go, I think. At least one who matters.'

'The leader?'

'Yes.'

'And how do you propose to find him?'

'I don't,' said James with a satisfied smile. 'He will find us. And I think it will be this weekend when a certain chess player arrives to pay court to Owyn's cousin.'

Gorath considered that, then nodded. 'He's a logical suspect, but how will you prove it? Accuse him in public?'

'Unlike your people, where I suspect an open challenge of honour carries some weight, this is a man whose honour is non-existent. He is one who lurks in shadows and kills from behind trees. He would only deny an accusation.'

'So then how do you get him to confess? Torture?'

James laughed. 'I've always considered torture to be of dubious benefit. Fanatics will die with a lie on their lips, and an innocent man will condemn himself to stop the pain.'

'I have found that torture, applied judiciously, can yield interesting results.'

'No doubt,' said James, with a look of mixed amusement and alarm.

Peter the Grey returned with his stable man and two workers, all of whom lost their sleepy slowness when they saw the bodies. 'Take them out back and burn the bodies,' instructed the innkeeper. As they complied, he looked at the shattered balcony railing and asked, 'Who will pay for this?'

James dug out a gold coin and said, 'I will. If I find the man behind this, I'll recover my gold from him. No need for you to bear the burden of the cost.'

'Thank you,' said Peter, greatly relieved.

Owyn returned with his uncle behind him, dressed in his

nightclothes with a large cloak around his shoulders. He was still barefoot. 'You've killed the Nighthawks?' he asked.

James said, 'I'm certain we've stamped out most of them in the area.'

Baron Corvallis was almost beside himself with glee. Then his mood turned darker. 'Most?'

'There's some business I think needs to be finished by Sixthday, then I think you'll be safe from the Guild of Assassins, m'lord.'

Corvallis said, 'Owyn, you couldn't have awakened me for better cause.' To James he said, 'I must pen a missive to Arutha, commending you to him for your good works this day.'

'Thank you, sir,' said James, 'but I'll be sending my own report to the Prince.'

'No false modesty, my boy.' He put a fatherly hand on James's shoulder. 'You must take praise where it comes. You might not be a squire all your life. Who knows, with a friend in court, and with recommendations such as mine, why some day you might rise to the rank of baronet or even baron!'

James grinned. 'One never knows.'

'Well, then,' said the Baron, turning toward the door. To Peter he said, 'Provide these gentlemen with whatever they need.' To Owyn he said, 'I can't tell you how pleased I am. I look forward to your company on Sixthday.'

He hurried out, and Owyn asked, 'What now?'

James looked at the mess and said, 'I think some sleep is in order.'

He retrieved his sword from where it had landed, cleaned it off on the tunic of the last dead Nighthawk, and as Peter the Grey returned to the commons, said, 'Master Grey, there's another dead one up in our room. Please remove it as well.'

'Oh, my word!' said the innkeeper.

*   *   *

'He's here,' said Owyn, hurrying into the room. Gorath and James had been resting on their beds, trying to relax after the fury of the night before.

James said, 'You're certain it's him?'

'Dandy, wearing fine clothing, and Ugyne is riding behind him with her head on his shoulder, just to annoy her father.'

'That's our man,' said James. 'Let him find us already half drunk.'

They hurried downstairs to an empty commons, and found things ready as James had requested. A chessboard had been set up and James had positioned the men as he wanted. Several empty tankards had been left nearby, and he signalled for Peter to bring over three half-filled.

Owyn sat opposite James and said, 'I hope you don't expect me to comment on this game. I have no idea what I'm looking at.'

'Good,' said James, 'because your part is to do nothing but look confused.'

Owyn's brow furrowed as he said, 'Well, I can do that with conviction.'

The door opened a short while later and Ugyne came in, almost skipping, leading by the hand a person who could only be Navon du Sandau. He was what James expected: tall, dressed in black with a white scarf around his neck. He wore a neatly trimmed pointed beard, a golden earring with a large diamond, and several golden chains which hung down his chest. He walked easily, with his left hand upon his sword hilt. James noted that while the hilt of the sword was decorative too, it was well worn, and the blade was almost certainly sharp and well-oiled. It was a rapier, and the only other man James knew who preferred the rapier as a weapon of choice was the Prince of Krondor. Light and agile, the rapier was a deadly weapon in the hands of a master, but in the hands of a novice, it was an easy way to get killed.

James had no doubt that Navon was a master. As Ugyne approached she said, 'Owyn, I have someone I want you to meet.'

Owyn looked up and said, 'Good. You can save me from humiliation.'

Ugyne introduced Owyn, James and Gorath, and said, 'This is my friend, Navon du Sandau.'

James nodded, doing his best imitation of a man who had started drinking early. He nodded slightly to Owyn who said, 'I think I should resign.'

With a smile, du Sandau said, 'Don't resign. Your position is difficult, but not hopeless.'

Owyn looked at James who again nodded slightly and Owyn said, 'Would you care to take over? I'm out of my depth.'

Navon said, 'If James doesn't mind?'

James shrugged. 'By all means. It was simply a friendly game; no stakes.'

Owyn stood up and stepped aside and Navon took his place. He studied the board and said, 'My move?'

James nodded. 'It's black's move.'

Navon studied the board and moved exactly as James had expected. James knew Navon was almost certainly a far better chess player than he was, but he had positioned the pieces as they had been during a game with the Keshian ambassador, Lord Abdur Rachman Memo Hazara-Khan, only he had been in Navon's position then. The ambassador had taken great pains to explain James's mistake to him after the match and the game was etched in James's memory. Navon had moved exactly as Lord Hazara-Khan had told James he should have moved in that long-ago game.

Ugyne showed Owyn a silver locket with a tiny emerald in it. 'See what Navon brought me?'

Owyn nodded appreciatively and watched the match. Both men took great pains to consider every option before they moved. After three moves James was convinced that should this game run its course, Navon would eventually win. Only by starting from a position of dominance was he able to appear competent enough to keep Navon's interest.

Gorath stood up, as if bored, and moved toward the door. 'I'll be back shortly,' he said to no one in particular.

This was Owyn's cue, and he said, 'Oh, Ugyne, do you remember that odd book on the family?'

'Which book?' asked the girl.

'The one with all those strange stories in it. You showed it to me when we were little. It was written by some cleric.'

'Oh!' she said, her eyes wide. 'You mean *The Abbot's Journal*! Yes, I do. It's funny, but I lent it to Navon here, a while ago, so he could learn about the family.'

Owyn said, 'Oh, I was hoping to read something in it I remembered from when I was a boy.'

James studied his opponent. If he was paying attention to the exchange behind him, he was a master of control. Not a twitch or flinch or even the slightest urge to turn and look at Owyn was evident. He was fixed upon the board before him.

Owyn asked, 'Navon, do you have the book with you?'

'What?' he asked, looking over. 'Book?'

'The family journal,' said Ugyne. 'I lent it to you a month ago.'

'Oh, that,' he said offhandedly. 'I left it at home. I'll return it next week.'

James nodded slightly, and Owyn returned the nod. He went to his backpack, which was on the floor behind Navon and withdrew the journal from the pack. He put the book upon the table next to the board.

Suddenly Navon rose, overturning the table as he did so, knocking James on his back. He threw an elbow at Owyn's chin, stunning the young magician.

Ugyne shrieked in alarm, and said, 'Navon! What is it?'

The man grabbed her by the wrist and turned her arm behind her back. He held her before him as he began backing toward the door. James came to his feet with his sword drawn, and saw Navon retreating. 'Stand back or I'll kill her,' he shouted, drawing his sword.

Ugyne shouted, 'You bastard!' and stepped down as hard as she could on his instep. While he hopped backward, she twisted away.

James reached out as quickly as possible and yanked the girl free, sending her sprawling toward Owyn, who caught her.

Navon glanced backward and said, 'I suppose your elf friend is standing outside the door.' He circled away from the door, putting his back to the wall.

James advanced, sword at the ready. 'Put that down and we'll have a chat. There are some questions that must be answered.'

Navon said, 'The instant I set eyes on you, I knew you were trouble. You look like that bastard Lysle Riggers down in Malac's Cross.'

James grinned. 'I've been told that before.'

Navon said, 'I assume you are the bunch who killed my men.'

'Sorry we couldn't accommodate them in their mission,' said James, 'but I have work yet unfinished.'

Navon leaped forward and lashed out with his blade and James parried. He knew he faced a master swordsman. The only comfort he took was that he had spent ten years practising with the best swordsman in the Kingdom. The exchange was quick: parry, counter, thrust and parry, and both men moved back.

'Well done,' said Navon, a note of honest appreciation in his voice. 'I don't suppose you could see your way clear to just backing away and letting me get to my horse.'

'Too many secrets, Navon. Or should I say Neville.'

Ugyne screeched, 'Neville!'

Navon's eyes widened slightly and a look of concern crossed his face. 'Say what you will, James of Krondor.

Soon it won't matter.' He launched another attack: a low, high, low combination that drove James back and almost got him killed as he tried to counter and Navon changed his line of attack.

James managed to avoid a lunge, getting inside Navon's extension and almost cutting him in return. After the two furious exchanges, both men stood dripping with perspiration, and knew they faced an accomplished opponent.

Owyn moved Ugyne away from the struggle, toward the kitchen, and said, 'Stay out of the way.'

'But your friend called him Neville. What is he saying?'

'What he's saying, dear sister,' said James's opponent, 'is you have been gulled into thinking I was dead.'

'Sister!' shrieked Ugyne, resisting Owyn's attempts to get her out of the way. 'My brother's dead!'

'I'll explain everything, after I kill your friend here.'

The fight continued. Every move was met by a counter, and every riposte was parried. The two men fell into a rhythm and each waited for the other to make a mistake. After another two minutes, James knew that's what it would come down to: whoever made the first mistake would die.

Back and forth they fenced, as fine a display of swordsmanship as had ever been seen in Cavell. Owyn tried to move to a place he might help James, but the movement of the two men was so precise and fluid, so quick and deadly, he hesitated lest he inadvertently cause his companion's death.

James's hair hung limply, drenched with sweat. He crouched low, sword ready, awaiting the next attack. The man known as Navon said, 'You're very good. Both chess and swordsmanship. A rare combination.'

'I had good teachers,' said James, using the pause to catch his breath. He studied every move of his opponent, waiting for some hint of what was to come next.

Navon stood motionless, also catching his breath. James was tempted to press the attack, then realized that was his opponent's tactic. As if to demonstrate the point, Navon let his sword point lower slightly, as if fatigue was making him sloppy. James calculated the odds of using this to his advantage. He said, 'I learned chess from the ambassador from Great Kesh.'

Navon smiled. 'Hazara-Khan! I would love to play him. I have heard he may be the best in the world.'

'Put down your sword, and I'll see if I can arrange a match. Of course, you'll have to play in the dungeon in Krondor,' and with that last word, James launched an intentionally poor attack, and as he suspected, Navon's response was fast and deadly. Only James's swift reflexes saved him.

Navon grinned. 'Close.'

'I've had closer,' said James, now sure of his opponent's abilities.

'Who taught you the sword?'

James started another bad attack, a high line with his sword hilt higher than the point, so that it appeared he was attempting to stab downward. Navon responded exactly as James had expected, and had James leaped back as most men would in that position, Navon would have skewered him. Instead, James leaned forward, his left hand touching the floor, allowing Navon's blade to pass over his back, actually cutting through the cloth of his tunic from shoulder down to mid-back. James rolled his wrist, bringing the point of his sword under, then up and Navon ran onto the point.

As the leader of the Nighthawks stood stunned in astonished silence, James said, 'I learned the sword from Prince Arutha.'

James pulled free the point of his sword and Navon collapsed to his knees. For a moment he stared at James with his eyes full of questions, but then life fled from them and he fell forward to strike the floor.

James put up his sword and knelt to examine Navon. 'He's dead,' said the squire.

Ugyne stood behind the bar, next to Peter the Grey, and demanded shrilly, 'What is going on?'

James stood up and said, 'We'll explain everything, but right now I need Owyn to go get your father. There's a mystery still to be unravelled.'

As Owyn ran to the door, James shouted, 'And watch out – '

Owyn opened the door, and Gorath unloaded a blow to the face that sent the young magician flying back into the room.

' – for Gorath,' finished James. He rose and crossed to where Owyn lay unconscious. Shaking his head, James turned to Ugyne and said, 'Could you please get your father, miss?'

The girl ran off to do as bid, and Peter the Grey came over and said, 'Pardon me, sir, but . . . well, I don't know any other way to say this: I really must ask you to leave.'

James looked at the mild-mannered innkeeper and laughed. 'I understand.'

A pale-faced Baron Corvallis arrived as they were hauling away the body of the man named Navon. James said, 'M'lord, we have a mystery to unravel.'

The Baron said, 'What is all this?'

Ugyne said, 'He called Navon "Neville", Father.'

If the Baron had looked wan when he arrived, what remaining colour had been in his face drained and he looked as if he might faint. 'Neville?'

James indicated the Baron should sit, and said, 'My lord, there's been murder done, not just recently, but years ago. Tell me about du Sandau and the wine cellar.'

The Baron put his hand over his eyes and leaned forward, and for a moment James thought he was weeping, but when he pulled his hand away at last, James saw mostly relief in his eyes. 'He was your brother, Ugyne. That is why I was so adamant about your not seeing him. He was courting you to enrage me.'

'I don't understand,' said the girl.

James said, 'Neville was your brother.' He looked at the Baron. 'But he was not your father's son.'

The Baron's colour rose and he nodded, looking as if he couldn't bring himself to speak.

James said to the girl, 'I did some snooping around. There are always those willing to gossip. It seems the man your father hired, Sandau, was a sculptor as well as a mason. He was reputed to have a way with the ladies. According to one of the old women I talked with, he was a big, handsome man, with a flamboyant nature, the type who appeals to some women.'

The Baron's face flushed.

Ugyne said, 'My mother was unfaithful?'

James said, 'It has been known to happen.'

She looked at her father as if he was a stranger. 'You had Sandau killed?'

'I arranged for an accident,' he said weakly. 'I didn't know it would get so out of hand. The cave-in killed a half-dozen men. And, I thought, Neville.' Looking as if he was growing angry, the Baron said, 'I didn't know the boy was going to be down there!' He slapped the table. 'I tried to treat him fairly.' Looking at Ugyne, he said, 'Your mother and I never talked about it after I found out. I tried to raise the boy as my own.'

She stood up and said, 'I don't know you.' She backed away a few steps. 'I don't know you at all.' She turned and ran from the inn.

James said, 'Baron, we have pressing business, but this will all be mentioned in my report to the Prince of Krondor. I suggest you take a trip to see your liege lord in Romney, and perhaps the King as well. To both of them you owe a complete confession and I think you need to put your affairs in order. I doubt the King will permit you back as Baron. I might also suggest you send Ugyne to stay with Owyn's family for a while.'

Owyn regained consciousness and said, 'What happened?'

Gorath helped him to his feet. 'I was expecting someone else. Sorry.' The last actually sounded sincere.

Owyn rubbed his swelling jaw. 'I'll be all right.' He looked around. 'What happened?'

'I'll tell you on the way.'

'On the way where?' asked Owyn.

James produced a key he had taken off of Navon. 'Back to Cavell Run.'

When the door was again open, James said, 'I knew that only a family member would be able to know how to trigger that door from the outside,' as Owyn jumped down from the ledge. 'If the other children in the village couldn't find it, Navon du Sandau from Kenting Rush wasn't going to blunder up here and find his way into the run.

'So, I asked a few questions and got the clues I needed,' he said as they walked back into the dark tunnel. Owyn produced another light with his magic and James continued. 'We've met the Baron. It doesn't take much imagination to see the Baroness attracted to a flamboyant, handsome man, even if he's a common mason. So Neville is conceived.

'The Baron finds out he's not the father, and he and his wife agree not to discuss it, but every day he sees the boy and is reminded of the betrayal.

'So, after a decade of daily insult to his manhood, he decides to lure the betrayer up to the run, rig an accident, and extract his revenge. Unfortunately, the boy was watching the work being done when the accident happened.'

Owyn said, 'And I wasn't here, and Ugyne couldn't open the entrance by herself from outside.'

'And perhaps the Baron himself didn't realize there was an outside trigger. I don't know and mostly I don't care. He killed at least four men and will have to be tried for that.'

They reached the barracks and headed for the stairs to the door with the lock. 'Neville somehow found his way out of the run. I suspect he was either injured, terrified, or both. We will never know how or why, but somehow he got out and lived. Someone found him and he survived. It might have been the Nighthawks, or he might have come to them later. It may be that a bright young and talented lad like Neville might have seized the opportunity to take control of the Nighthawks when Arutha all but stamped them out in Krondor, ten years ago. It would have been the right time for survivors of that destruction to have been seeking sanctuary in an out-of-the-way place like the run.

'They changed his appearance enough so it wasn't apparent to those who knew the boy that he was the same person. Some people change dramatically between eleven summers and twenty-two. Or maybe they used some magic. As I've said, we'll never know. But we do know there were relationships that Neville inherited, between the Nighthawks, the moredhel and the Pantathians.'

Gorath almost spat at the last. 'Damn snakes and their hot land magic. I can't abide them.'

'But Murmandamus counted them useful,' said James, not knowing that Murmandamus had actually been a Pantathian in disguise, magically altered to look like a moredhel.

He reached the lock and used the key he had taken from Navon. The key turned and the lock opened, and James pushed up on the trap door. It swung away and he mounted the steps and found himself in private quarters. He quickly glanced through the single door and found another barracks, empty and unused for years. But the small room contained chests with gold, gems and documents.

James ignored the gems and gold, and quickly read through documents.

After a moment, he said, 'Damn me!'

'What?' asked Owyn.

'Northwarden. Delekhan is attacking through Northwarden.'

Gorath said, 'Why?'

James was silent for a moment, holding up his hand to fend off questions while he thought. 'It makes sense. That's why all this murderous insanity has been under way.

'If Delekhan overruns Northwarden, he can come down the River Vosna; it runs along the northern foothills of the Calara Mountains, and runs through Mastak Gorge. From there it's only a short portage to the headwaters of the River Rom. After that, he's only days from Romney. Romney!' He looked at Owyn and Gorath. 'That's why all the troubles in Romney. He needs a city in chaos so it can't mount resistance.'

'Why Romney?' asked Owyn.

'Because from there he takes the River Rom south and where it turns back toward the southeast he lands and marches overland to Sethanon. There's nothing but open plain and light woodlands by that route.'

Owyn said, 'And by burning the keep at Cavell and occupying the run – '

' – he prevents anyone from occupying a strong position behind his lines,' finished James.

He stood up and hurried down the steps. 'We must leave now.'

Gorath and Owyn hurried after. 'Where are we going?'

'I'm heading for Northwarden,' answered James, 'to warn Baron Gabot of the attack. You need to take these documents to Arutha.' He handed three rolled-up parchments to Owyn.

'Arutha?' Owyn shook his head. 'Unless we use your Tsurani orb, it'll take us weeks to return to Krondor.'

'He's not in Krondor, so the orb is of no use,' said James as they reached the waterfall exit. 'He's encamped within the northern edge of the Dimwood with a large portion of his army, waiting for word on where the attack is staging, so he can rush to support. He can be within sight of Tyr-Sog, Highcastle, or Northwarden within a week of getting word.'

'So you want us to tell him to come to Northwarden.'

'Yes,' said James, as he scrambled down wet rocks to where the horses were tied.

'What if he doesn't believe us?' asked Gorath. 'He seemed dubious about my claims when last we met.'

'Far less dubious than he appeared,' said James. 'Let me advise you never to play cards with the Prince. In any event, if he expresses doubts, tell him, "There's a Party at Mother's". That way he'll know the message is from me.'

Owyn said, 'Odd, but we will.'

'James,' said Gorath, 'if the Prince is in the Dimwood, so will be the advanced elements of Delekhan's forces. If the final goal is Sethanon, many of my people will have filtered down through the small gullies and passes in the Teeth of the World and will be readying things for the advancing army next spring.'

'Well do I know,' said James. 'I remember when we evacuated Highcastle and rode across the High Wold and down through the Dimwood.'

'What if we're captured or killed?'

Mounting his horse, James said, 'I have one thing to say to that.'

'What?' asked Owyn.

'Don't be,' said James, turning his horse and riding off.

Owyn mounted and said, 'Let's stop so I can see Ugyne safely on her way to my parents, and we'll get some food.'

Gorath said, 'That would be wise.'

Owyn said, 'Then that's about the only wise thing about this plan.'

ELEVEN

Escape

A pebble clattered down the hillside.

Gorath had his sword in hand before it stopped rolling, and said, 'Owyn!'

The young man from Timons stood peering into the night, blind from having gazed at the campfire. From out of the darkness a voice spoke in a language Owyn didn't comprehend. Arrows slammed into the dirt at Owyn's feet, to emphasize whatever command was given.

Gorath said, 'Don't resist. We're surrounded.'

A group of men and moredhel advanced into the light. One of them walked up to Gorath and looked him in the eyes a moment; then with as powerful a blow as he could muster, he struck Gorath across the face. Owyn looked at the moredhel, sure he had seen him before, but not certain where.

Then the moredhel advanced upon Owyn, and spoke the King's Tongue. 'You must have conspired with that walking garbage to kill my brother.' Suddenly pain exploded in Owyn's face as he was struck.

In shock and dizzy from the blow, Owyn lay on the ground. He realized that this must be the brother of the magician Nago, whom they had slain in Yellow Mule. To the two of them, Narab said, 'I would happily put your head on a pike, human, and hoist it while I drag this traitor behind me from here to Sar-Sargoth, but I am going to give that pleasure to Delekhan.' Turning to the others, he said, 'Drug them, bind them, and bring their horses!'

Owyn was roughly pulled upright and a bitter drink was

forced past his lips. He tried to spit it out and was hit hard across the face for his trouble. His head was cruelly pulled back and his nose held while the concoction was poured down his throat. He was forced to swallow. A few moments after he had, he felt his legs and arms growing leaden, his mind confused, and his vision hard to focus. He found his hands tied tightly behind him and a blindfold tied around his head. Then he was hoisted into his saddle by rough hands. Once there, his feet were lashed to his stirrups, and the horse was led away. Other men and dark elves appeared, leading horses, and Narab ordered them to mount.

The nightmare ride began.

The horses were changed many times, and Owyn remembered resting for a period – minutes or hours he couldn't recall – but he knew time was passing. The drug was obviously designed to dull his mind so that whatever magic he might have possessed was unavailable to him. Several times he became aware enough to realize the drug was wearing off, but then he was given more to drink. Once he fell awkwardly from the saddle and hung by the ties on his feet, forcing his captors to halt and right him. They added more ropes.

He was vaguely aware of being thirsty and hungry, but it was a distant discomfort. Mostly he existed in a grey fog, punctuated by the constant pounding from the horse upon which he rode. Then he was dragged from the horse and hauled through a cold, damp place and cast down onto rough stones. He lay there for a time, lapsing in and out of consciousness. Then, eventually, one moment ceased passing into the next, and he awoke in pain. He moved slowly, and discovered himself free of leg restraints, though his arms were still bound and he was still blindfolded.

Owyn sat up and moved his aching and stiff legs. The insides of both of them were bruised and he knew he had ridden a long way without being able to sit a comfortable seat. Even had he been conscious he sensed the ride would have been punishing; it had taken at least seven or eight days, from what he could recall, and he had switched horses a number of times. But with senses dulled and tied to his saddle, it was only by the gods' mercy he was still alive.

The sound of footfalls, heavy boots on stone, approached and the sound of a cell being unlocked announced the arrival of his captives. Hands yanked Owyn to his feet and he couldn't avoid groaning in pain.

The blindfold was removed from his eyes and even the relatively low brightness of a torch outside the cell caused Owyn to blink. A dagger cut through the ropes around his wrists, and when he moved his arms, agony shot through his shoulders. The pain almost caused him to fall, but he was held upright by two guards.

Narab came to stand before Owyn and said, 'He should still have enough of the drug in him to be harmless.' He turned and they escorted Owyn out of the cell. From a cell next door, Gorath was also escorted, and Owyn noticed he didn't seem to be in better shape, though he walked with apparently less discomfort.

The tunnel was long and dark, and Owyn sensed it was deep underground. Despite his dulled magic senses, he knew immediately that at one time great power had resided here. There was something ancient and terrible about this place, and despite his drug-dulled senses, he was very afraid.

They were taken through a series of tunnels to a landing from which rose broad stairs. They were escorted up the stairs along a wide hallway, and led to a massive chamber. In the centre of the chamber rested a massive throne, currently empty. At its right was another, smaller throne, upon which sat a large, powerfully-built moredhel, who could only be Delekhan.

Narab said, 'Master, I have a prize for you.'

The guards pushed Owyn and Gorath forward, so they

landed sprawling at Delekhan's feet. 'What is this?' Delekhan demanded, rising to stand over Gorath.

'Gorath of the Ardanien! I have captured him. Let me have the honour of cutting out his heart to revenge my brother's death.'

'Your brother was a fool!' shouted Delekhan. Owyn looked up at the towering figure, and saw a broad face, surprisingly blunt of features for one of elvenkind. His face was a mask of rage, the most expression Owyn had seen on a dark elf so far. 'And you are as well,' Delekhan added. 'You've wrecked everything, you dog!'

Owyn looked at Narab, who stood white-faced, almost trembling with shock and outrage. 'But . . . I have brought back a traitor! We can torture him to discover the names of the other dissidents – '

'You know *nothing*!' Delekhan turned to the guards. 'Return those two to their cells below. I will question them later.' To Narab he said, 'Your life hangs by the slimmest thread. Presume one more time and your head will adorn a pike outside the gate!' Walking away, he said, 'Now get out of my sight, you bungler, and do not dare to approach me until I send for you.'

Although Owyn was no expert on the facial expressions of the moredhel, he could see murder clearly written on Narab's face. And it was directed at Delekhan's retreating back. Owyn was jerked around by two guards, hauling him to his feet, and once again he was forced to march back into the bowels of the dungeons at Sar-Sargoth.

No food or water was brought, and Owyn considered it academic, as they were likely to be dead within hours. Time passed slowly and Gorath was silent. Owyn felt no impulse to speak, as he was awash with numbing fatigue. The ride, the lack of food and sleep, the drug, all were making it difficult for him to do anything but lie on the icy stones and attempt to rest.

Time passed slowly, a blur of thoughts which fled before they were remembered, perhaps he dozed for a while.

Then suddenly he sat up, his skin tingling. Magic! Energized by the fey effect of someone, somewhere casting an enchantment, he reached for the bars of his cell. A metallic click sounded and the bars pushed open. 'Gorath!' he said in a harsh whisper.

Gorath looked over and his eyes widened as he saw Owyn free. 'Someone is using magic to set us free!' Owyn said, moving through the door, his injuries and fatigue forgotten.

Gorath tested his door and found it also unlatched. 'Who?' he wondered.

'I have no idea,' said Owyn. 'Whoever helped you escape the north the last time, perhaps?'

'Let us worry about that later,' advised Gorath. 'We must get out of this fortress before we are missed.'

They moved through the halls of the dungeon. At the large hall that led upward they found a dead guard, his blood freshly pooled on the floor. 'Whoever threw the spell must have done it from here,' suggested Owyn.

'Over there,' said Gorath, pointing to a table upon which were piled the belongings that had been stripped from the two prisoners. Gorath put on his sword and tossed Owyn his staff. Owyn said, 'I don't suppose they left me any of my gold?'

Gorath said, 'Hardly.'

Owyn knelt and examined the dead guard. He came away with a small pouch. 'Well, this will have to do.'

Moving to the stairway, Owyn asked, 'Do you know a way out of here?'

'Several,' replied Gorath. 'This city was built for tens of thousands of my people to occupy. If Delekhan has more than a few hundred outside of the central palace area, I'll be shocked. Moreover, many of the tribes here are strangers to one another, and there are many human renegades as well, so once we are free of the central palace, we may be able to use guile to find our way out.' He moved up the stairs. 'But only if we are away from here when they find we are gone.'

Gorath led Owyn up a flight of stone steps, through a hall, and down a dark passage. Moment to moment they expected to hear the alarm raised behind them, but no hue sounded.

Suddenly they were above ground, in a courtyard devoid of life. Gorath motioned and Owyn followed, the twin spurs of fear and hope moving him despite his injuries and the drugs still in him.

They hid in a grove of scrub as fresh snow fell. 'Does spring ever visit this land?' asked Owyn.

'Yes,' said Gorath slowly. 'Very late, and our warm days are too few. But yes, we do see spring.'

'I thought Yabon a cold place,' said Owyn.

'What is your home like?' asked Gorath.

'Timons? Warm, most of the time.' Owyn stared into the distance. 'We get rain quite a bit and occasionally great storms off the sea, but in the summer it's quite hot. My mother tends the garden and my father breeds horses. I didn't realize how much I miss it until now.'

'Why did you leave?'

Owyn shrugged. 'A boy's foolishness. My father had a servant, a magician from the north named Patrus who lived with us for a time. He taught me my first lessons. After I studied a while at Stardock, I came to understand that he wasn't very powerful as magicians go, but he was *very* smart. He understood things. I think that's what I was really looking for, how to understand the world better.'

Gorath was silent a while, then at last he said, 'I think we all would be better off if more of us sought understanding and fewer of us sought power.' He glanced at the fading light. 'Come, it is time.'

They had been waiting for darkness, to attempt to slip out of the precinct around the fortress. Moredhel warriors and renegade humans, infantry and mounted soldiers had been moving for hours. At first they had assumed they were the object of a search, but after a while it was clear this was far more than a hunt for a pair of fugitives. This was a mobilization.

Gorath led them through a series of snow-filled gullies, over a hill, and then down a long draw that led to a flat plain south of the city. 'The plain of Sar-Sargoth,' said Gorath. 'Legend has it this is where the Valheru met in council. Great circles of dragons rested there while their riders assembled.'

Owyn saw a sea of tents and a large pavilion in the centre, in front of which rose a standard: a crimson field upon which a white leopard crouched. 'How do we get around that camp?'

'We don't,' said Gorath, leading him toward the centre of the encampment. 'If we don't find friends here, at least I think we shall not find enemies.'

Several moredhel warriors glanced at Gorath and Owyn as they walked through the camp. They appeared indifferent to Gorath and Owyn's approach, though one got up and ran ahead. By the time they reached the large pavilion, the occupant stood waiting at the door to greet them.

'Greetings, Gorath of the Ardanien. Were not the dungeons of Sar-Sargoth to your liking?' The speaker was a striking female moredhel. Tall and regal of features, her hair was gathered into a knot behind her and allowed to fall in a cascade of dark red. She wore armour in the same fashion as the males of her tribe, yet even in her warrior's garb, Owyn was struck by her beauty. Alien and strange it was, but no less compelling for that. She stepped aside, indicating that they might enter. She waved them to a place near a small fire. 'Eat, rest for a while. I thought Delekhan would have killed you by now. Your escape will cause him no little discomfort.'

'You sound pleased at the prospect, Liallan.'

'My husband's rise took me with it, Gorath,' she said, 'but our marriage had nothing to do with affection. It was a wedding of powerful tribes, to seize control of our respective clans, and to keep them from shedding one another's blood . . . for a time. Nothing more.'

'Is that why the charade, Liallan? You don't believe in Delekhan's mad plans any more than I, yet you openly support him. You command a tribe as powerful as his own; your influence in council is second to none but Narab.'

'You've been gone from us too long, Gorath. Much has changed in a short while. Narab even now musters his clan, and turns to face Delekhan.' She sat down next to Gorath and took a small piece of meat from a simmering pot next to the fire. She placed it between Gorath's teeth in a gesture that was clearly seductive, yet even Owyn could tell it was a ritual rather than an open invitation. 'Our new master is displeased with Narab. Something to do with your capture, I believe.'

Gorath accepted the ritual offer of food, then handed a bowl to Owyn. Owyn tore off a large piece of bread from a loaf next to the plate and used it to scoop up a mouthful of hot stew. Gorath said, 'Why would your husband be upset with my capture? He certainly tried hard enough to keep me from fleeing south.'

Liallan sat back. She looked at Gorath for a moment, then said, 'You are a warrior of great honour, Gorath, and your bravery is unquestioned, as well as your caretaking of your clan, but you are naive at times.'

Gorath looked ready to take umbrage and studied the woman with a narrow gaze. 'You come close to giving insult.'

'Don't take it as such. In these cynical days, your openness and honesty are refreshing.' She reached up and unbuckled her breastplate, removing her armour. Owyn saw she wore a simple sleeveless tunic beneath the armour. She possessed a long neck and slender arms, yet there was nothing frail about her. Her movements hinted at speed, and the muscles of her arms and neck showed power. She was a dangerous woman, by any race's measure.

'What are you saying, Liallan?'

'I'm saying you were picked for a role. You were the ideal clan chieftain for this part.'

'I was *allowed* to escape?'

Liallan said, 'Who do you think engineered your escape from Sar-Sargoth all those months ago?' After a moment, she said, 'I did. Just as I misdirected Delekhan's soldiers into the snow plains while Obkhar's family fled to the mountains near the Lake of the Sky. If they avoided the eledhel and the dwarves at Stone Mountain, they may be safely back in the Green Heart.'

'Why?' asked Gorath.

'To keep Delekhan busy,' said Liallan. 'He has his timetable, I have mine. It suits my purpose to delay his assault of the Kingdom a while longer. His stupidity in treating with Narab will buy me another month. Once Narab's head is upon a pole at Sar-Sargoth's gate, it will take at least a month for Delekhan to bully the fractious clan leaders back into obeying him without question. Delekhan wants an early-spring campaign; I prefer one a little later in the year.'

Owyn asked, 'Did you help us escape?'

'This time? No,' said Liallan. 'I reap no gain in doing so. Whatever you may have done, you achieved on your own.'

Owyn said, 'No, someone else opened that cage.'

'Then I suspect it may have been Narab. That fit of pique is what I would expect of him. If Delekhan threatens him for capturing you, then why not release you?'

'Will you help us again?' asked Gorath.

'I will consider such an effort an investment against the future of the Northlands, Gorath. Killing you or turning you over to my husband gains me nothing. Letting you go costs me little, and in the future your help may be useful. I have agents throughout the Northlands, and I will send word to certain of them to aid your travels south.'

Gorath said, 'I will do what I can to assist you if fate allows.'

She smiled, revealing perfect white teeth. 'Rest for a while, then I will have horses ready for you. Take to the west, and avoid the roads. The best route is by what the humans call the Inclindel Gap, south of Sar-Isbandia. But avoid the village of Harlik, for Moraeulf camps there and he knows you well.'

She stretched and Owyn was again struck by her beauty and catlike grace. 'Rest now, for in the morning things will become quite lively outside the city. Narab's clan answers his call, and Delekhan will no doubt call down the wrath of the Six upon him. It should be over shortly.'

'Who are these six magicians?' asked Gorath.

Liallan's voice dropped to a near-whisper, as if someone might be listening. 'They advise, and more. They scheme into the night with Delekhan. Only a few see them, and no one knows who they truly are. It was they who advised Delekhan to obliterate your tribe.'

'But why?' demanded Gorath. 'We were never among Delekhan's rivals, even if we served only with reluctance.'

'Because you were small, and your tribe had long been one to stay aloof. When your father died, you took your people and fled to the cold northern mountains. Wise, but it made you suspect. You avenged yourself, which was expected, but among those you killed were those related to Delekhan by blood. He could not ignore your acts, for he was under scrutiny, and he was driven by his need for powerful allies. In short, you made a bitter enemy, and your tribe's destruction was an effective object lesson. As will be Narab's death.'

'Did the Six order that?'

Liallan shrugged. 'I do not know, but I would not be

surprised if my husband didn't hear warnings over the last few months casting doubts on Narab and Nago. Your slaying of Nago did Delekhan a favour. He was reluctant to move against one brother while the other was alive. Together, they were the two most powerful spell-casters of our nation, and their clan is not one that can be ignored.'

Gorath ate in silence a moment, then said, 'Where did the Six come from?'

'No one knows. No one even knows what race claim them. They are Spellweavers far beyond the powers of our race. Some suspect they may be Pantathians come among us again.'

'Murmandamus,' Gorath said softly.

'Yes,' said Liallan. 'The same as those who served the Marked One.'

'Do they abide in Sar-Sargoth?'

'When they counsel Delekhan. Presently they are with his son Moraeulf in Harlik. They seek out more fugitives from your clan, those who are trying to win freedom and get south to the Green Heart.'

Gorath said, 'Then I have even more pressing reasons to carry warning to Prince Arutha. If I cannot get my hands around Delekhan's throat, I will aid one who will bring him low.'

'Tread carefully,' said Liallan.

To Owyn it sounded as if she were being sincere in her concern.

'Perhaps all our schemes will bear fruit. If I raise my Snow Leopard banner above the walls of Sar-Sargoth, Gorath, you and the surviving Ardanien will be welcome to return to the heart of their people.'

Gorath's expression was guarded. 'You are as much to be feared as Delekhan, Liallan.'

She smiled and again looked dangerous. 'Only by those who seek to harm me or my tribe, Gorath. Return to your northern mountains in peace if that day comes.' She stood, and said, 'Rest. I will have horses outside before sunrise.' As she reached the doorway, she looked over her shoulder and said, 'Hide well and move quickly, Gorath. If you return to my sight before Delekhan is overthrown, I must needs present him with your head as a peace offering.'

'I understand, Liallan. You've been generous to one humbled by fate.'

She left, and Gorath said, 'She's right, Owyn. We need to rest.'

Owyn lay down next to the fire, content with a full stomach, and glad to be rid of the drugs that had dulled his senses for so many days. Still, it seemed as if only a moment passed between closing his eyes and Gorath's shaking him, saying, 'It's time.'

He rose and forced stiff aching muscles to obey as he wrapped a heavy fur-lined moredhel cloak around him and mounted a waiting horse. If the guards were curious as to who Liallan's guests were, they said nothing, merely standing aside as the two strangers rode off.

The building was run-down, but there were a dozen horses tied in front of it. 'We can get something to eat inside,' said Gorath.

The purse Owyn had liberated contained a few coins, Kingdom, Quegan and even a Keshian silver piece, as well as some gems. They dismounted and Owyn said, 'What is this place?'

'You'd call it an inn. One of the conventions brought to the north by your people. My kind have never created such, but we have come to appreciate their benefits.'

They went inside to find a dark, small room, with as many as twenty men and moredhel standing around. A bar that was little more than long planks set upon barrels ran along the far wall of the building. Gorath shoved aside two men and said, 'Ale and something to eat.'

The human barkeeper produced a platter of cheese and

bread, surprisingly good given the shoddy surroundings. They ate, and Owyn trusted Gorath's instincts on his ability to blend in. 'Where are we?' he whispered.

'Near the City of Sar-Isbandia. What you humans call Armengar. There are villages and towns throughout this region. Much trading with the south.'

Owyn said, 'Most of us who live in the Kingdom think of the Teeth of the World as a wall separating our peoples.'

'It's a barrier to warfare, perhaps, but enterprising men find a way to trade. There are a dozen ways through the mountains south of here.'

From behind, a voice spoke lowly. 'And all are heavily guarded, Gorath.'

Gorath spun, his hand falling upon his sword hilt. 'Draw steel and die,' said the other moredhel. 'Eat your cheese and live.'

Gorath didn't smile, but his face relaxed. 'I see you've managed to keep your head attached to your shoulders, Irmelyn.'

'No thanks to Delekhan,' said the other moredhel. He indicated with a nod they should move to a small table in the corner. Owyn picked up the cheese, took his ale and followed.

Sitting in the crowded room, the moredhel named Irmelyn said, 'Delekhan will have the rivers running piss and chickens laying dust by the time this all ends. Drink while you can, my old foe.'

'Why are you here, Irmelyn? I was told Obkhar's tribe had fled.'

'Most have, but a few of us remain behind, in the hope we can free our chieftain.'

Lowering his voice to a whisper, Gorath asked, 'He's alive?'

Irmelyn nodded. 'He's alive, and close by. He's being held prisoner in the naphtha mines under the destroyed city.'

'Prisoner?' Gorath looked confused. 'Why isn't he dead?'

'Because Delekhan doesn't know he's working as a slave in the mines. They think he is a man called Okabun, from Liallan's Snow Leopards.'

Gorath said, 'So you linger nearby to free him?'

Irmelyn nodded. 'We do. We need help. Would you care to provide that help?'

'In exchange for what?'

'For a way south. As I said, the passes are all heavily guarded, but I know a way to get through.'

'What do we need to do?' asked Gorath.

'Come outside.'

They rose and left the relative warmth of the inn. Once they were outside Irmelyn said, 'We have discovered a way out of the mines. Unguarded.'

'Then why doesn't Obkhar just walk out?' asked Owyn.

With a snarl, Irmelyn said, 'When I want to hear from you, pup, I'll kick you.'

Gorath said, 'Then tell me, why doesn't Obkhar just walk out?'

'Because of the fumes that hang in the tunnels. When the humans fled after firing the city, several tunnels from the old keep collapsed. One didn't, but it is small, and the fumes that hang there would explode if a spark was struck. They would overcome anyone seeking to pass.'

'But you have a plan?' said Gorath.

'We have found masks, used by humans in the old days, constructed of bone and membrane from a dragon's lungs. They let air pass through but keep the deadly fumes out.'

'So you need someone to get inside and get a mask to Obkhar,' said Owyn.

The tall moredhel glared at the young human, but said, 'Yes, we need someone to get a mask to Obkhar and escape with him.'

'Why us?' asked Gorath. 'Why not a member of your clan?'

'There are only a few of us left in the Northlands, and

Moraeulf's soldiers know us all. You, on the other hand, while known by name, are not well known by sight. The Ardanien lived apart for many years; you could claim to be a member of any number of clans and who would say no?'

'What do you propose?' asked Gorath.

'Go to the slaver, a human named Venutrier. He claims to be from the Kingdom city of Lan, but I know him to be a Quegan. Tell him you wish to sell the boy.'

'What?' Owyn was about to object.

Gorath held up his hand. 'Say on.'

'Venutrier is as venal a human as you could wish to meet. He will certainly try to capture you. Let him.

'Two of his guards will be alerted and allow you to enter the mines with your bundles and will store them for you. When you are taken below, they will come to you with your bundles and leave you unwatched. Obkhar will be somewhere on the level to the west of the great gallery. More than that we can't tell you. If you agree and get him out, we will see you and your companion safely south.'

Gorath said, 'Before I say yes or no, tell me: have you word of Cullich?'

Irmelyn said, 'Yes, she is not far from here. A hut between here and the village of Karne. We can see her on our way south if that is your desire.'

Gorath was quiet for a moment, then said, 'It is. We will do it.'

Irmelyn said, 'Then walk to the mine entrance. You will be challenged. Tell the guard you wish to speak to Venutrier. I will take your horses and weapons and meet you at a place Obkhar knows.'

'Care to tell us?' asked Owyn.

'If you do not free Obkhar, you have not kept your part of the bargain, human. You can fare as well as you may without our aid.'

Gorath said, 'Come along, Owyn. We have a distance to walk.'

Without looking back, he led the human away and set out for the mines.

Venutrier was a huge man, gross fat barely contained by a massive belt he wore around his waist. He looked over at Owyn and said, 'Where'd you catch him?'

'I didn't,' said Gorath. 'He's a runaway kitchen whelp from the Kingdom who thought to come fight for gold. Well, he couldn't play knucklebones and it turns out he can't pay his gambling debts.'

'He's a bit scrawny,' said the slaver. 'Come with me.' Without waiting to see if Gorath followed, he walked toward the mine entrance.

They entered the mine and Venutrier asked, 'Who are you, warrior?'

'I am Gorath of . . . the Balakhar, from the Green Heart.'

'Not from around here?' said Venutrier. 'Good. We could use a strapping worker such as yourself.'

Guards lowered spears and suddenly Gorath and Owyn were surrounded. 'Had you been from here, my friend, you would have known that no one comes without allies to my mines. Lord Delekhan has ordered an impossible amount of naphtha for the invasion of the Kingdom and I need workers. Get them below.'

Gorath and Owyn were hustled below by the guards and taken to the second level of the mines, as Irmelyn had predicted. Then they were taken to a large empty cavern.

One of the guards lingered as the others walked away, and he whispered, 'Stay here.'

They remained alone for a period, the darkness cut through by only one faint light, a lantern cleverly fashioned with a thin transparent membrane covering the flame. 'I don't expect we're going to see a lot of torches around here,' observed Owyn.

'If there are fumes of naphtha in the tunnels, I expect you're correct.'

Shortly a guard returned, carrying the bundles taken from Owyn and Gorath. He also carried a third bundle. 'Here. Take that tunnel there. You will be facing west. Find your friend and then go down to where you hear water. You must swim out.'

The guard vanished and Gorath picked up the new bundle. It contained three odd-looking devices, obviously designed to wear over the nose and mouth. They gathered up their remaining possessions and departed.

The tunnel to the west went downhill, and abruptly Gorath stopped.

'What is it?' asked Owyn.

'We must be under the old city of Sar-Isbandia.'

Owyn didn't know what to say.

Gorath continued walking. Soon they came to a large gallery, where the sound of work could be heard. A single guard moved idly around the huge gallery, overseeing the wretches labouring to lift buckets of the thick oil that ran through the earth, to bubble up to the surface.

Owyn's eyes teared and he said, 'I can see why they need the mask if it gets much worse than this.'

Gorath said, 'Look for one of my people who wears his hair in a high fall, and who has a scar running down his face from forehead to chin.'

When the guard was at the farthest point in his rounds, they slipped through the main gallery to another tunnel. Those who laboured hardly spared them a glance, intent as they were upon their own miseries.

Not seeing Obkhar, Gorath said, 'Let us continue to the west.'

They moved down a long corridor that turned into another gallery, and in that one laboured a small band of moredhel.

Owyn looked around and said, 'I don't see any guards.'

Wiping away tears, Gorath said, 'I think they linger near the fresher air at the ends of the tunnels. Where would these prisoners flee to?'

'Nowhere, Gorath,' came a voice from behind them.

They spun to be confronted by a large, gaunt moredhel who possessed the scar Gorath had described. 'Obkhar!'

Looking Gorath up and down, Obkhar said, 'At first I thought the fumes had finally taken my senses, but I see they have not. How is it you are here? I heard that your head had been spitted on a stake outside Sar-Sargoth.'

Gorath folded his arms across his chest. 'Not all who remain in the Northlands willingly bend to Delekhan's will. And not all who rebel die. I had help in escaping, as you do now. Others died so that I might win free.'

'You have a grave debt to repay.'

'All the more reason to see Delekhan's reign ended, Obkhar! He shall pay blood debt to me and mine.'

'Most of my kin are now in the Green Heart, but should you raise your banner against Delekhan, Gorath, we will come to your cause.'

Gorath smiled. 'So you at last forgive me for giving you that scar?'

Laughing, Obkhar said, 'Never. I still intend to kill you for that, some day, but for the time being we need to be allies.'

Owyn produced the masks. 'Where is the tunnel of fumes?'

'This way,' said Obkhar, leading them down a side tunnel.

They reached a point where the fumes threatened to suffocate them, and Obkhar said, 'Put on your mask. They will help your eyes not at all, but you will be able to breathe. We have a long way to go and an icy swim at the end of it. The tunnel out is half flooded, and leads to a branch of the River Isbandi.'

They put on the masks and Owyn was surprised to discover they worked. The fumes burned his eyes, but by blinking rapidly he could see. He almost gave Obkhar a heart attack when he illuminated himself and his companions with his magic. The old moredhel chieftain said, 'For a moment I thought you had struck a flame, and we were all about to be incinerated.'

They reached the tunnel that was flooded and entered icy water that rose to their knees. As they walked they moved deeper and soon they were up to their chests. Obkhar signalled and ducked his head underwater. Owyn and Gorath did likewise. They felt a tug and suddenly were swept into an underground stream.

Kicking hard, Owyn followed and when he came up, his head bumped stone. Fighting down panic, he moved a short distance away and his head broke clear of water. Obkhar said, 'You can take your mask off.'

'Good,' replied Owyn. 'Because mine came off underwater.'

Something that may have been a chuckle came from Gorath. Obkhar said, 'We have less than a mile to swim.'

They set off, Owyn fearing he would be pulled down by the weight of his sodden clothing, but he mustered the strength to continue. Suddenly above he saw stars and he realized they had come outside.

A short way down the river torches burned and when they swam toward them, voices softly called out.

'It is I, Irmelyn.'

They were helped out of the water, taken to a fire and given heavy robes to wear while their clothes were dried. 'Any alarm?' asked Obkhar.

'None so far,' said a moredhel unknown to Owyn. 'The guards we bribed will say nothing. It may go unnoticed for a very long time that you are not there. Many die in the mines and their bodies lie unnoticed in tunnels.'

Gorath asked, 'Now, what of Cullich?'

Obkhar said, 'Is she still alive?'

Irmelyn said, 'Yes, and she lives nearby.'

'I was told I could see her on our way south,' said Gorath.

Obkhar looked at Irmelyn who nodded.

'A promise is a promise,' said the chieftain. 'I must leave now, with those of my tribe who are to travel the passes with me. Irmelyn will guide you to Cullich and then on your way over the mountains.'

'Avoid Harlik,' said Irmelyn. 'Moraeulf and the Six are there.'

'I will,' said Obkhar, as he finished changing into dry clothing. He said: 'Gorath, fare you well, old foe. Let no one but me take your life.'

'You survive,' said Gorath, 'so that I may take your head some day.'

After they had gone, Owyn said, 'You two sound almost fond of one another.'

Gorath ate a piece of dried beef given him by Irmelyn and said, 'Of course. Friends can betray you, but with an old enemy, you always know where you stand.'

Owyn said, 'I never thought of it that way.'

Irmelyn said, 'They are an odd race, aren't they?'

'Very odd,' agreed Gorath.

The hut was primitive, barely four walls of scrap wood cobbled together and roofed with thatch. A stone chimney emitted a faint wisp of smoke, the only sign of anyone inside.

'She's in there?' asked Gorath.

Irmelyn nodded. 'Yes.'

Gorath dismounted, as did Owyn. Irmelyn said, 'Delekhan has her watched occasionally. I had better stay here. If I call, come quickly.'

Gorath nodded, and opened the door.

If the woman who waited inside was shocked at the unexpected appearance, she masked it well. She merely looked up from her corner next to the fire and said, 'Enter and close the door.'

'Is that your warmest welcome, Cullich? Your husband has returned.'

Owyn's mouth dropped open.

She rose, sinuous and powerful in her movement, and while her gown was in tatters and her hair dirty and matted, Owyn was struck by the resemblance between this woman and Liallan. Despite the fact that this woman's hair was raven dark, and Liallan's red; and where Liallan had been slender and lithe, Cullich was buxom and broad of hip, her face wide-boned, there was something in common with the sunken-cheeked leader of the Snow Leopard Clan. Both women radiated power.

'Husband?' said the woman in mocking tones, her blue eyes fastened on Gorath. 'How so? Clan leader? By what right? Ruler of a host? No more. Once you held those titles and had earned that rank, with guile and bravery, cunning and strength. Around you the Clan Ardanien lay curled like a sleeping dragon, awaiting your word to rise up and crush whoever opposed us. Where is that dragon now?'

'Gone, scattered to the north, across the Teeth of the World, hiding.'

'Then call yourself clan chief and husband no more, Gorath. You lost the right to those titles when you gave the order to flee Sethanon and refused my wisdom.'

'Wisdom, old witch? You counsel murder and madness. Do you still dream of conquest, of all the ranting of Murmandamus? Did you learn nothing by the obliteration of our people at Armengar and Sethanon? Two sons did I see fall along the way. One of them was *our* son.'

'What would you have of me, old man?' asked the woman.

'I seek to end the madness. Will you aid me?'

'How, by dying and having my head placed on a spear outside Sar-Sargoth?'

'Delekhan must be stopped.'

'Why? What destiny would you choose for our people, Gorath? Would you have us bend our heads to the earth once more? Should we serve the eledhel Queen as we once did the Valheru? We are a free people! Or do you feel the tug of the Returning?'

'No!' said Gorath, his eyes flashing in anger. 'But I have heard things, learned things.' Pointing to Owyn, he said, 'Not all humans are our enemies.'

'No,' said Cullich. 'There are those who will serve us for gold.'

'No, I mean there are those who would live with us as neighbours, in peace.'

'Peace?' said the woman, with a laugh of contempt. 'When have the moredhel spoken of peace? You sound like one returned to Elvandar. They who were once rampaging bulls are now gelded oxen, serving the Queen, no better than slaves.'

'This is not so, wife,' said Gorath. 'The glamredhel have joined the eledhel, and not as slaves, but as welcomed brethren.'

'The mad ones!' said the moredhel woman. 'You think it true, then you go. I will abide. Here is my home, and eventually I will find someone who can use my talents and my knowledge, and he will be a warrior, and I will show him how to rise and take power and how to hold it. I will have other sons, sons that will live.'

Gorath sighed. 'I feared that such would be your reply.'

'Then why have you come here? Surely not to rekindle a love long dead between us.'

'No . . . I need your help. For a short time, then I shall be gone from your life, one way or another.'

'For the sake of that love, now dead, I will listen,' she said, openly surprised by Gorath's admission.

'Where are Delekhan's forces now?'

Cullich looked out of the frosted window. 'Massed on the Kingdom border. The banners of Clans Krieda, Dargelas and

Oeirdu are held in reserve near Raglam. I hear both Liallan and Narab's forces are to march soon.'

Gorath smiled. 'Narab has turned on his master, like a rabid wolf.'

'Nevertheless, there are ample armies along the border to make crossing difficult.'

'We have a way,' said Gorath.

'Then what would you have of me?'

'You know things, witch. What do you know of the Six?'

'I once sought to scry on them, and for my troubles I was rendered senseless for more than a day. I know only that they possess arts beyond my understanding. Of all the things Delekhan has his hand in, this may be the most dangerous. He thinks he controls them; I wonder.'

From outside the house, Irmelyn shouted.

'We must leave!'

'Go,' said the witch. 'I think we shall never see one another again, and for that I am not sad. Too much pain has passed between us. These will be our last words as husband and wife. When you pass through that door, our marriage will end. But know this: I wish you well in whatever life awaits you.'

'As do I,' said Gorath sadly. 'Be well, wife.'

'And you, husband.'

Gorath left the hut and when the door slammed shut, ending his marriage, he hesitated for an instant, then he and Owyn mounted and rode off.

Irmelyn shouted as they rode: 'We must clear a pass before sundown, or those who will look the other way when we go by will have been replaced.'

Lost in thought, Gorath said nothing, and Owyn could only think that with luck, he might live to see the Kingdom again.

TWELVE

Preparations

Wind and rain pelted the riders.

Owyn wasn't sure if this was preferable to the snows he had endured on the north side of the mountain, for while it was warmer, it was far wetter. His heavy fur-lined robe was sodden, weighing on him like lead. But at least, he thought, this time he wasn't drugged and tied to his horse.

The escort provided by Obkhar's clan had seen them safely to a pass controlled by their faction. As they reached the foothills of the mountain, they intercepted a runner carrying warnings of a falling-out near Sar-Sargoth. Delekhan's forces were surrounded by Narab, who had been removed from Delekhan's inner council and replaced by Delekhan's son, Moraeulf. Speculation was that Narab had to move to capture and destroy Delekhan before the Six intervened or else he and his clan would be crushed. Gorath greeted the news with indifference, later mentioning to Owyn that he would be pleased if either of them destroyed the other.

At the summit of the small pass they had taken, the escort turned back, saying this pass was heavily patrolled by Kingdom forces. As if predicted, later that same day they had been intercepted by a Kingdom patrol of Krondorian regulars. The officer in command, Lieutenant Flynn by name, had been ready to brand them both renegades, but Owyn mentioned Arutha's name and said they carried a message from Squire James; and more to the point, they knew Arutha was camped in the Dimwood.

The patrol had handed off Gorath and Owyn to another detachment, who had escorted them to a camp in the

Dimwood. For several miles, the bivouacked soldiers' fires were visible. Gorath had observed that a significant portion of the Kingdom army must be in the woods.

Arutha sat at a command table, Knight-Marshal Gardan at his side, looking at marks on a large map of the mountains leading to the north. Looking up as Gorath and Owyn were ushered into his presence, he said, 'You look on the verge of collapse. Sit down.' He indicated a pair of camp chairs nearby. Owyn didn't need a second invitation and sat heavily, while Gorath walked to the map and studied it. 'Here,' he said, putting his finger on the spot designated Northwarden. 'This is where Delekhan plans to assault your forces.'

Arutha was silent for a long while, studying the moredhel. Finally, he said, 'If you will forgive my caution, where is Squire James?'

Owyn said, 'Sire, he sent us to bring you word while he hurried to Northwarden to carry warning to Baron Gabot. He gave us these documents.' He handed the documents to a soldier who gave them to Knight-Marshal Gardan. Owyn filled them in on how they had uncovered and destroyed the nest of Nighthawks near Cavell Keep. He detailed James's theory that Delekhan was planning on going by boat and portage from Northwarden to Romney, then straight overland to Sethanon.

Arutha again was silent as he studied the documents. 'These are much like those we saw when first you came to Krondor, Gorath. Then they claimed the attacks were in places like Tannerus and Yabon. What are we to believe?'

Gardan's dark face was set in an expression of doubt. He said, 'We hear you speak of leaving James at Cavell Keep, yet we intercept you coming south through the mountains again. You picked a most indirect route to reach us, moredhel.'

'We had little choice, my lord,' replied Owyn. He explained about the capture and attempted to outline the chaotic conditions among the various clans of the north.

When he finished, Arutha said, 'You paint a picture of confusion and rival factions battling for control, yet our patrols and advanced units see only a unified opposition, working in a co-ordinated fashion.'

'You see only those forces loyal to Delekhan south of the Teeth of the World, Prince Arutha,' said Gorath. 'Clans who either oppose or resist him are either fleeing to refuges in the ice-bound mountains to the far north or seeking to travel near the Lake of the Sky south past the eledhel and dwarves to the Green Heart.'

Gardan said, 'We have had reports from Duke Martin of heavier than usual sightings of bands of moredhel moving past the eastern boundary of Crydee, Highness. Martin says he's seen women and children, so they're not war parties.'

Arutha said, 'I am still dubious. I sent Locklear two weeks ago to gather reports from the border barons to the east. He is going to Highcastle and Northwarden. He should return in another two weeks. If James is there, Locklear will return with word.'

Gorath said, 'James said you might need to be convinced. He said to tell you . . .' He glanced at Owyn.

'There's a Party at Mother's,' said Owyn.

Arutha nodded. '"And a good time will be had by all." It's a Mockers' password, used by James and me the first time we met.'

'Do you believe us now, Highness?' asked Owyn.

'I believe that James believes this to be true,' said Arutha. He sat back thinking. 'I just hope he's right.'

'Orders, Highness?' asked Gardan.

'I have no choice. Either I trust James's intelligence or I don't. I want a detachment left behind to secure this area, but the balance of the army is to march to Northwarden.'

Gardan studied the map. 'Would it not be wiser to alert the King and muster the Army of the East to reinforce Gabot?'

'It would if the Army of the East was mustered already. I'll send a message to Lyam asking him to be ready to stand behind us, should Delekhan win past Northwarden. But we can be there faster than Lyam, so let us be expedient. Order camp broken at first light tomorrow.'

Gardan saluted and left the tent to give orders. Arutha said, 'Tell me about the Six.'

Owyn tried to recall everything that was said about the mysterious magicians working for Delekhan. When at last he had finished, prodded by several acute questions from the Prince, Arutha said, 'I have a mission for you two.'

Gorath said, 'I would rather be on the wall at Northwarden, Highness, so that I might greet Delekhan as he deserves.'

'I have no doubt,' said Arutha. 'But personal honour and debts of blood must be put aside. If we all fail, who will revenge us? I want you to go back to Krondor, to find Pug. If he is not there and his wife Katala is, she will be able to reach him. If she has also gone, simply use a talisman Pug gave me for the purpose. The Princess knows it and how to use it, and when Pug comes, tell him of the Six. I think magic will play an even bigger part in this coming conflict, and I am ill-prepared if we are to encounter such at Northwarden.'

'Cannot the boy alone do this?' asked Gorath.

'Pug will have means to extract things from your memory you may have forgotten,' said Arutha. 'But I doubt he can do such without your help.'

Gorath was silent for a long while, then said, 'Once this is done, I wish to return and fight.'

Arutha nodded. 'I understand.' Then he paused. 'No, I don't understand. That was presumptuous. I know nothing of your race and what drives you.' He studied Gorath's face for a moment, as if trying to read something inside the moredhel chieftain. 'But I would like the opportunity to learn some time. I can appreciate the drive to right a wrong, personally. When you are finished with Pug, return and I will welcome your sword.'

Gorath said, 'You are also more than I expected, Prince Arutha. I also would appreciate the opportunity to learn more of your people.' He glanced at Owyn. 'Though this boy and the other have shown me a great deal already that has made me question many of my people's attitudes toward your race.'

Arutha said, 'That is a beginning. Perhaps one day we can have more.' He came around the table and extended his hand to Gorath, who took it. They shook hands and it was more than a gesture.

'Your Highness is gracious,' said Gorath.

'Rest, and tomorrow go with the patrol I send to Malac's Cross. It is faster than trying to go straight through the woods toward Sethanon and around the mountains to Darkmoor. I'll have documents drawn and you can commandeer an escort at Malac's Cross and at Darkmoor. They should get you to Krondor safely. Once there, Pug will know what to do.'

Owyn and Gorath departed, and a soldier escorted them to a tent. He held aside the tent flap and said, 'The lads who sleep here are on patrol until tomorrow, so they won't mind your sleeping here if you don't steal nothing.' He smiled to show he was joking, but Gorath fixed him with a stare that caused the smile to fade. He hurried away saying, 'There's food at the big fire near the Prince's tent when you're hungry.'

Gorath said, 'It will be good to eat hot food again.' He glanced over to one of the bedrolls to find Owyn already face-down and snoring.

James cursed all petty barons who answered only to the King as he negotiated his way along a frozen ridge, his breath forming clouds of white before him as he exhaled. The air stung each time he inhaled, his toes were numb, and his stomach reminded him he had not eaten yet.

James had arrived within hours of Locklear at Baron Gabot's fortress, a towering keep of stone which dominated one of the three major passes through the eastern half of the Teeth of the World. Unlike Highcastle, which had sat in the middle of the pass itself, providing a barrier that was a controlled gate, Northwarden rose up on a small peak, around which wound the pass known as Northland's Door. A single road wound down the side of the large hill in a lazy s-curve, widening as it descended. Designed this way, the road gave the double benefit of allowing the Baron's forces to spread out as they charged down to intercept any foe, while forcing any attackers to concentrate a smaller force in the van should they be foolish enough to attack up the road.

What kept the road below in Baron Gabot's control was a series of siege engines mounted on two walls, the north and the west. The western defences were the heaviest, while the northern were designed to harry any forces attempting to come down the pass and negotiate the turn up the road to the keep. Mangonels and catapults, as well as a trio of heavy ballistae over the main gate, ensured that any army attempting to pass would take critical casualties before they rounded the pass and got beyond the engines' range. Some soldiers would get past, it was certain, but nothing resembling an organized force. And to deal with any who did win through, the Baron kept a small garrison of horse soldiers in a barracks near the small town of Dencamp-on-the-Teeth.

Baron Gabot had felt confident that any threat coming through Northwarden could be dealt with by his command. That had been a welcome response to James, though he hoped fervently that Owyn and Gorath had reached Arutha in the Dimwood and help was on the way. He was beginning to worry. Had they reached Arutha and convinced him of the warning, the Prince's army should have been arriving at Northwarden now.

Instead, there was only silence. Gabot had sent another message to the Dimwood, at James's urging, requesting support from the Prince, and had also sent word south via fast messenger to the King, his liege lord. At least, thought James, Gabot wasn't as stiff-necked as old Baron

Brian Highcastle, who had managed to get himself killed ignoring Arutha's advice when Murmandamus had driven south over his position. With luck, Arutha would receive Gabot's message even if Gorath and Owyn hadn't survived.

James found himself hoping that wasn't the case; he had grown fond of the youth from Timons, and he was surprised to find he also had come to like something about the moredhel. He couldn't quite put his finger on it, but there was something definite about the dark elf, a lack of uncertainty about who or what he was; few men had it, and James admired it: even more, he admired the moredhel's ability to put aside his own personal dislike for humans to seek their aid in opposing what he saw as a great wrong against his people.

Locklear waved and pointed. As a favour for Baron Gabot, since dawn, he and James had been scouting ahead to see if advance moredhel units were anywhere in the north end of the pass. A patrol had headed out two days before, accompanied by a magician now in the Baron's employ, and the Baron was concerned about their fate. It went unsaid that the two squires were no loss to the Baron should any harm befall them, while losing another patrol to the enemy would severely weaken Northwarden. James and Locklear couldn't contrive a plausible reason to say no, so here on the second day of their trip they were working their way through the frozen dawn, with James silently cursing all border barons.

A noise ahead had alerted them to a possible enemy position. Locklear was holding his horse while James climbed above the floor of the pass to a high ridge to get a look ahead. A single figure scampered along the trail, holding the hem of his ivory-coloured robe with one hand, exposing spindly legs as he hurried. In his other hand he held a large staff, shod at either end with iron caps.

Every hundred feet or so, he would turn and pause, and when a pursuing figure would come into view, he'd unleash a bolt of energy, a blast of flame the size of a melon; a tactic that was producing little real damage, but which served to keep the pursuer from closing. James began scrambling down the hillside, while Locklear shouted, 'What is it?'

Sliding the last dozen yards, James hit the ground running and said, 'I think wc've found Gabot's magician.' He pulled a crossbow off the back of his horse and quickly cranked it up and placed a bolt in it, while Locklear drew his sword and waited.

The old man rounded a corner and hesitated when he saw the two squires. Locklear signed for him to come on, and shouted, 'This way!'

The old man hurried and when the moredhel who was chasing him rounded the same corner, James drew a bead on him, then let fly with his crossbow. The bolt sped across the gap and took the moredhel right off his feet, propelling him backward.

Locklear said, 'You've been practising. I'm impressed.'

'I'll never learn to use the bow, but this thing is pretty easy,' said James, putting away the crossbow.

'Not very accurate, though.'

James nodded. 'Find a good one, then keep it. Some of them shoot all over the place; this one usually hits what I'm aiming at.'

The old man was puffing a bit and when he reached them, he put his staff down and leaned on it. 'Thanks, lads. That was a little closer than I care to think about.'

'Are you Master Patrus?' asked Locklear.

'Just Patrus,' said the old man. 'Yes, I'm he. Why, you looking for me?'

James said, 'And a company of Baron Gabot's soldiers.'

The old man was slender and sported a wispy grey moustache and goatee. He wore a hat that looked more like a nightcap than any sort of proper hat, and along with the ivory-coloured robe, it made him appear to be walking about in his nightclothes. Pointing back the way he had come, he said, 'We got jumped a half-day back, by a mixed company of those damned Dark Brothers and trolls. Those trolls were a handful, I can tell you.'

James said, 'I've fought them. You're the only one to get away?'

'One or another of the lads may have found a way through. Some of them got up into the ridges. I'm an old man; best I could do was hurry along the road and keep them ducking behind me.'

'Where did they jump you?'

'About two miles ahead,' said the old magician.

'That staff of yours is handy,' observed Locklear.

'Well, lad, the truth is it's a little bit of fire, not much more than a scorch mark if it hits you, but it's just hot enough to make you duck if you see the fireball coming at you. I made the thing years ago to impress some pesky townspeople down south who were trying to run me off. A few little fireballs tossed their way and they left me alone.'

James laughed. 'Owyn didn't tell me you were such a character.'

'Owyn Belefote? Where do you know that rascal from?' asked Patrus.

'It's a long story. I'll tell you while we walk. If you're up to it, I want to check out the place those trolls jumped you. Otherwise you can continue back to Northwarden. It should be safe between here and there.'

'I think I'll stick close to you, lads. Who are you?'

'I'm Squire James of Krondor, and this is Squire Locklear. We're members of the Prince's court.' They started walking their horses rather than ride while the old man walked.

'Prince Arutha's lads? You wouldn't happen to know Pug of Stardock, would you?'

'We've had the pleasure,' said Locklear.

'I'd like to meet him, some time. I've heard a thing or two about his academy. Told Owyn he ought to get himself down there; I'd taught him everything he could learn.'

James said, 'Locklear here met Owyn on his way back from Stardock; he was visiting his aunt in Yabon. I think Stardock didn't work out too well for him.'

'Bah!' said the old magician, picking his way along the road with his staff. 'The boy has talent, a fair amount from what I can tell, but I think he's one of those Greater Paths, because a lot of what I tried to teach him just didn't work. But the things that did, why, he was fierce with it, he was.' The old country magician looked up the pass and said, 'Company's coming.'

Locklear drew his sword and James unlimbered his cross-bow again. But rather than trolls or dark elves, two dusty members of Baron Gabot's company came into view. One was obviously wounded and the other looked very tired.

'Patrus!' said the wounded soldier. 'We thought they'd got you.'

'Not even close,' said the old man with a grin. 'These lads lent a hand.'

'I'm Squire James. What did you see?'

The senior-most soldier reported, indicating that a squad of twenty Dark Brothers and an equal number of trolls had ambushed their patrol, and only a falling-out between the two factions had kept them from killing all of Gabot's men.

'That's interesting,' said Locklear.

'Very,' agreed James. 'If they're fighting, it's over pay.'

Patrus nodded. 'Troll mercenaries don't wait to get paid. They go back home or take it out of your hide.'

'I don't know what caused the row,' said the wounded soldier, 'but we were running and one of the Brother-hood of the Dark Path yelled something at a troll, and instead of chasing us, the troll turned and tried to slice up the Brother. It was a fair mêlée by the time we got away.'

The other soldier nodded. 'They had their blood up, the trolls did, and they seemed just as satisfied killing Dark Brothers as they did us.'

'Great,' said James. 'Confusion to the enemy. Now, you boys all right to get back to the Baron alone?'

'If there's no one waiting between here and there to jump us, we'll be okay,' said the wounded soldier.

'Good. Go and report to the Baron and when you're done telling him what you've seen, tell him we're going to go snoop around and see what else we can find.'

'Very well, squire,' said the unwounded man, saluting.

The soldiers continued on and Locklear said, 'What do you have in mind?'

'If those soldiers got jumped by trolls, there's a camp nearby.'

'Yes,' said Patrus. 'The town of Raglam's ahead. It's sort of an open town. Not quite Kingdom, but enough humans living there that it's not particularly Northlands, either. Lots of weapons runners, slavers and other no-accounts visit there all the time.'

'Sounds like my kind of place,' said James with a grin.

'You going to get us killed?'

James's grin widened. 'Never, Locklear, my old friend; you're going to get killed some day over a woman, not because of anything I'm planning.'

Locklear returned the grin. 'Well, if she's beautiful enough.'

They laughed, and Patrus said, 'You boys got something you'd like to tell an old conjurer like me about?'

'I thought we might take a ride into Raglam and have a look around.'

Patrus shook his head. 'Crazy, that's what you two are. Sounds like fun.'

The old magician started to march up the draw, and James and Locklear exchanged glances, then laughter.

The patrol leader signalled for his men to halt and said to Gorath and Owyn, 'Malac's Cross.'

They were arrayed before The Queen's Row Tavern,

which was obviously crowded, and Owyn said, 'Why don't we try the abbey?'

Gorath nodded. They bid their escort goodbye and rode on, and Gorath said, 'I would have thought you'd prefer an ale and the company of others than the monks of Ishap.'

'I would, had I the means to pay for that ale,' said Owyn. 'Unless you've secreted away some booty you failed to mention to me, I'm without a copper to my name, thanks to Delekhan's guards. In all the preparation for heading off to Northwarden, the Prince was so busy . . . I forgot to ask for funds.'

Gorath said, 'So we beg?'

'We ask for hospitality. I suspect Abbot Graves is a more likely source for such than an overworked innkeeper.'

Gorath said, 'Perhaps you're right.'

'Besides, we might even convince the Abbot to lend us the price of a meal or two between here and Krondor.'

'We should have thought of that before leaving Prince Arutha.'

'I didn't think of it,' said Owyn. 'You didn't think of it. We didn't think of it. So, there's no "should", is there?'

Gorath grumbled that this was so.

They reached the abbey and saw that the gate was closed. 'Hello, the abbey!' called Owyn.

'Who is it?' came a voice from within.

'Owyn Belefote. We came to see the Abbot.'

'Wait,' was the terse reply. And they waited.

Nearly a quarter of an hour passed before the gate opened, and a very worried-looking monk admitted them. As soon as they had passed through the gate, it slammed behind them. Dismounting, Owyn asked, 'What is this?'

A monk took their horses and said, 'The Abbot waits for you within.'

They went inside and found Abbot Graves overseeing a pair of monks who appeared to be packing things up.

'Are you leaving?' asked Gorath.

Looking at the two, Graves said, 'Where is James?'

'Last we saw him he was on his way to Northwarden,' replied Owyn. 'Why?'

'Damn!' swore the Abbot. 'I was hoping he could do me a service.'

Owyn repeated Gorath. 'Are you leaving?'

'I must,' said Graves. 'Twice in the last week Nighthawks have tried to kill me.'

Owyn and Gorath exchanged questioning looks. Owyn said, 'But Abbot, James killed the leader of the Nighthawks.'

'Navon is dead?' asked Graves.

Before anyone could react, Gorath had his sword drawn and the point levelled at the Abbot's throat. Two monks leaped to their feet, one trying to put as much distance between himself and the moredhel as possible, while the other assumed a fighting stance, as if ready to defend the abbey's leader.

'Wait!' shouted Owyn, putting his hands out.

'How did you know du Sandau was the leader of the Nighthawks?' demanded Gorath. 'We could have been killed for lack of that knowledge.'

Graves held up his hands. 'Because he was extorting me.'

Owyn put his hand on Gorath's sword and slowly forced the point down. 'Let's talk,' he said calmly.

Graves motioned for the monk who was ready to attack to withdraw and the young cleric nodded and departed, the other monk a step behind him.

Gorath said, 'Explain this "extorting" before I kill you.'

Owyn said, 'Sandau was forcing Graves to do something against his will by threatening him with something. Isn't that right?'

'Yes,' Graves replied. 'He found out something about me and used it to gain my help in whatever he was plotting.'

Owyn sat on the table where the monks had been working and said, 'How can anyone force a priest of Ishap to do anything? You have magic and a powerful church to call on. What did he do?'

Graves said, 'As I told Jimmy – James, I have ties from my old life that aren't completely severed.' Graves sat and Gorath put his sword away. 'I used to be a thief, a basher, for the Mockers in Krondor. I provided protection for cargo we were running in and out of the city, and kept anyone else from setting up a gang, and I protected our girls, so no one roughed them up.'

He looked down and his expression was one of regret. 'When I felt the call and went to the Temple of Ishap, I tried to put that life behind me. The church trained me for two years, and I took vows. But I wasn't honest in my vow.'

'How could you lie taking a vow in a temple?' asked Owyn, his expression showing astonishment. 'It can't be done!'

Graves said, 'It can, if you don't know it's a lie when you make it. I honestly thought I was rid of my past, but I was lying to myself.'

'What does that mean?' asked Gorath.

'I thought I had severed all ties, but I hadn't. When I was placed in the brotherhood of monks, I was asked to work on behalf of the temple in Krondor. So I was back among my old haunts.'

He fell silent, as if reluctant to go on.

'What happened?' asked Owyn.

'There's a woman. She was a girl I knew when I was a basher. She was as tough as a boot and mean as an alley cat. That's what we called her, Kat. Her name is Katherine.'

'A whore?' asked Gorath.

'No, a thief,' said Graves. 'She was a fair pickpocket and tough enough to be a basher, but where she really excelled was boosting. She could steal your nightshirt off you while you slept and you'd wake naked and wondering where your laundry was.' He sighed. 'She was a little slip of a thing when I met her. I used to tease her and watch her get mad at me. Then when she got older I'd tease her and she'd tease back.

'Then I fell in love with her.'

'But you left her to take orders with Ishap?' said Owyn.

'She's a lively thing, and she could do better than me. A lot of the younger boys would like to take up with her. I thought she would be better off with someone else. I thought it would be easy to put her out of my mind. But it wasn't.

'I saw her on the streets from time to time, and somewhere, somehow, a fellow named the Crawler got wind of her, and one night this Navon du Sandau comes up to me, bright as gold and sits down at a table at the Queen's Cross and says, "We know about your little kitty cat in Krondor. If you don't do what we tell you to do, she's dead." He said if I asked the temple for help she'd be dead.'

'You believed him?' asked Owyn.

'I had to. He knew things. This Crawler had been looking for people for a long time, I guess, because he knew enough about my old life I knew he'd kill her before I could do anything.'

Owyn said, 'So why are you getting ready for travel?'

'I was expecting a message a month ago from du Sandau. Instead a Nighthawk tried to climb the wall of the abbey. The brother responsible for defending the abbey intercepted him and it was close, but the assassin died.

'Then two weeks later, I was walking back from the centre of town when a crossbow bolt intended for me struck the brother walking next to me.'

'Where are you going?' asked Owyn.

Graves said, 'I am owed some favours by a man in a village near Sloop. He has dealings in Kesh. I sent him a message asking him to help me get out of the Kingdom. Today a message came from him indicating he could help.'

Owyn said, 'Michael Waylander?'

'Yes,' replied Graves. 'How did you know?'

Owyn said, 'There is a relationship here. Waylander, you,

the Nighthawks, and this Crawler. I'm not sure I can begin to guess at it, but if James were here he might puzzle this out.'

'I can't wait. Even if Sandau is dead, there are other Nighthawks. The one who shot at me is still out there.'

'True,' said Gorath, 'but won't your order protect you?'

Graves shook his head and his expression was one of regret. 'If I had gone to them at once, perhaps. But I didn't, and I've broken my vow. My only hope is to get Kat out of Krondor, and to reach Kesh before the Nighthawks find me.'

'We're heading for Krondor,' said Owyn. 'Should we travel together?'

'Your magic and your friend's sword would count for a lot, but you'd be putting yourself in harm's way.'

Owyn laughed. 'I've been doing that on a regular basis since I met Gorath.'

'Life is danger,' Gorath said. 'I do not understand how your love for this girl could blind you to your duty, but then much about you humans is strange to me. If Owyn says we should not kill you for your part in this business with the Nighthawks, I will follow his lead.' He leaned forward, his boot on the bench on which Graves sat, until his face was before the Abbot's. 'But if you betray us again, I will eat your heart.'

Graves smiled back, and the old basher could briefly be seen, as he said, 'You're welcome to try at any time, elf.'

Gorath snorted.

Owyn said, 'Well, we are lacking funds, so we must needs depend on your generosity to eat on the road.'

Graves stood up and called for his monks, who returned to help him finish packing. 'If you get me to Krondor alive, you'll have earned your meals and some gold as bonus.'

Owyn said, 'If that Nighthawk is out there watching the abbey, he'll know we're here.'

'We leave tonight,' said Graves.

Owyn winced. 'I wanted to sleep in a bed,' he complained.

'Sleep now,' said Graves, pointing to his own pallet in the corner of the room. 'I'll wake you when it's time to go.'

Owyn nodded. 'If we must.'

'We must,' said Graves.

Owyn lay down on the straw-stuffed mattress on the floor, and Graves said to Gorath, 'Would you like to sleep?'

'Yes,' replied the dark elf, but he remained standing, his eyes on Graves. 'But *after* we're on our way to Krondor.'

Graves nodded and returned to overseeing the preparations for his departure.

THIRTEEN

Betrayal

The trolls looked up.

James said, 'Just keep moving slowly, like we know what we're doing.'

Patrus whispered, 'Do we know what we're doing?'

'Don't ask,' Locklear replied.

The trolls were raising weapons and spreading out to fight. James slowed his horse and said, 'Just keep moving, but be ready.'

The trolls were roughly human in appearance, with almost no necks. Their heads thrust forward from their shoulders, so they always looked as if they were shrugging. James knew their somewhat comical appearance was as far from the truth as it could be. The lowland trolls were little more than beasts, without language or the ability to use tools and weapons. Their mountain cousins were intelligent, if stupid by human standards, and knew how to use weapons. Very well. Their language sounded like grunts and squawks to humans, but they had a social organization and knew how to fight.

As the trolls approached, James held up his hand in greeting. 'Where is Narab?' he asked conversationally.

The trolls halted their advance, and looked one to another. They had low foreheads and jutting lower jaws and large teeth, with two lower tusks that protruded up over their upper lips a short way. One turned his head as if listening and said, 'No Narab here. Who you?'

'We're mercenaries, but we've been sent to find Narab and find out why you trolls haven't been paid.'

At the mention of payment, the trolls began an excited conversation. After a few minutes, the first troll to speak – James assumed he was the leader – said, 'We no fight if we not paid.'

'That's the problem,' said James. He leaned over the neck of his horse and spoke conversationally. 'Look, I understand. If I were you and weren't getting paid I wouldn't fight either. I might even just take my lads and go home, the way this Delekhan's been treating you.'

'You pay?' asked the troll, holding his war club in a suddenly menacing fashion.

James quickly sat back in his saddle, ready to spin his horse away if he saw that weapon moving with any but the most casual purpose. 'I suppose,' said James. He turned to Locklear and said, 'How much gold do you have?'

'My travel allowance!' hissed Locklear. 'A bit more than a hundred good sovereigns.'

James smiled. 'Give it to them.'

'What?'

'Just do it!' insisted the senior squire.

Locklear took off his belt pouch and tossed it to the troll, who caught it with surprising dexterity. 'What this?'

'A hundred golden sovereigns,' said James.

'Gold is good,' said the troll. 'We work for you now.'

James grinned. 'Very good; then stay here until we get back. And if anyone is following us, stop them.'

The troll nodded and waved his companions aside so that James could pass. As they moved away from the trolls Locklear said, 'Why don't we just buy them all off and send them home?'

James said, 'Truth to tell, it would be cheaper in the long run. But the dark elves are unlikely to set so low a price.'

Patrus said, 'Mountain trolls are only one thing more than stupid, boys.'

'What?' asked Locklear.

'They're greedy. You think that bunch is going to let us just ride past and not ask for more?'

'No,' said James, 'which is why I have this other purse here, in case they do.'

Locklear said, 'So that's why you needed my gold? So you could use your own on the way back.'

'No,' said James. 'If we can get back without paying, we will. I had you use your gold, because I didn't want to give them *my* gold.'

Locklear snorted and Patrus laughed. They moved along the road and after a while saw a company of riders moving at a leisurely pace along the horizon. 'We must be getting close,' said James.

'Yes, Raglam's just on the other side of that rise,' said Patrus.

They plodded along, attempting to look unconcerned and relaxed as they rode into the heart of enemy territory. James had managed many times in his young life to go places he wasn't supposed to be simply on the strength of looking like he knew where he was going and had a reason for being there, and he hoped that proved as truc with dark elves as it had with humans.

They rounded a corner as they topped the rise, and James halted. 'Gods of mercy!' he exclaimed.

Engineers were hard at work building siege towers for the walls of Northwarden. 'Well,' said Locklear, 'I don't think we have to see much more to convince the Baron they are coming this way, do we?'

Patrus walked forward. 'Let's see what else they're up to.'

They passed a bored-looking band of humans, sitting alongside a huge catapult. A moredhel warrior walked toward them. 'Where are you going?' he demanded.

James assumed a look of indifference. 'Where's Shupik?'

The moredhel said, 'Who?'

'Shupik. Our captain. We're supposed to report to him, but we can't find his camp.'

'I have never heard of this Shupik,' said the moredhel.

Before James said anything, Patrus said, 'It's not our fault you're ignorant, you pointy-eared lily-eater! Get out of our way so we can find our captain, or you can explain to your chieftain why he didn't get the information we were sent to fetch back here!'

Patrus set off at a brisk walk and James and Locklear moved after him. James gave the moredhel a shrug as he walked past. As they rode on, Locklear muttered, 'And I thought *you* were brazen.'

James could barely suppress a laugh. They passed half a dozen towers under construction and James said, 'Someone did their fieldwork. Those will be hard to get up the road to the keep, but if they can move them quickly enough and they reach the wall, they'll fit snug up there and get warriors on the wall in quick order.'

Locklear nodded. 'Nothing like those big lumbering monstrosities at Armengar.'

James nodded. He remembered the huge war engines being pulled across the plains of Sar-Isbandia to the walls of Armengar. Only the brilliance of Guy du Bas-Tyra had kept those machines from reaching the walls time after time. James doubted Baron Gabot would prove as able a defensive general.

As they rode past, Locklear said, 'Some shallow trenches on the road a half mile or so before the walls might cause them some problems.'

James grinned. 'Serious problems, especially if we started throwing things down onto the road.'

'Like boulders?' asked Patrus, who then began to laugh, a sound that could only be called 'evil'.

Locklear was openly cheerful as he said, 'Could be quite a mess.'

As they moved down the road, Locklear said, 'Say, Patrus, how did you end up here in the middle of this?'

The old magician shrugged. 'Old Earl Belefote ran me out

of Timons for "infecting" his son, as he called it. Like the boy wouldn't have discovered he had talents without me. Anyway, I wandered a while, up to Salador, where that Duke Laurie was downright hospitable to magicians. But I get bored easily if I don't have something to occupy myself with, and Laurie said that Gabot had wanted someone up here who knew about magic to advise him about these Dark Brother Spellweavers, so I came up and have been working with the Baron for the last year or so.'

'What have you discovered about the moredhel Spell-weavers?' asked James.

'Got some notes back at Northwarden. A lot of little things. Not much that makes sense, at least as I understand magic. I wish I knew more about the elves out in the west, then I might have a better idea about what I've learned. When we get back to the castle, I'll show you what I've come up with. But right now,' he said, pointing ahead, 'I think we have a problem.'

James slowed down as they approached two bands of warriors, humans on one side and a mix of humans and moredhel on the other. They were involved in a heated exchange and by the time James and his companions reached them, they appeared to be on the verge of open conflict.

'I don't care what he says,' exclaimed the apparent spokesman for the human-only faction. 'Kroldech isn't fit to command fleas attacking a dog.'

'You're bound by oath! You took gold, human!' retorted a moredhel war chieftain. 'You'll go where you're ordered, or you'll be branded traitor.'

'I signed on with Moraeulf! I took *his* gold. Where is he?'

'Moraeulf serves his father, Delekhan, as we all do. Moraeulf is in the west, because his father wills it. If Delekhan places Kroldech at our head, then that is who we're following.'

James appeared uninterested as they rode by, but he listened to every word.

When they were a short distance past, Locklear said, 'Dissent in the ranks.'

'Pity,' said James, dryly.

James reined in.

'What is it?' asked Locklear.

'Look at that catapult.'

Locklear looked at the war engine. 'What about it?'

'Does something about it strike you as funny?'

'Not particularly,' said Locklear.

Patrus laughed. 'You'll never make general, boy. If you were to move that thing, what would you do first?'

Locklear said, 'Well, I'd unload it – ' Suddenly Locklear's eyes widened. 'It's loaded?'

'That's what your sharp-eyed friend was trying to make you see,' said Patrus. 'Not only is it loaded, it's pointed the wrong way.'

'And unless I'm mistaken, that rather large rock in the basket end of the arm is sighted to land right over there on that inn.' James moved his horse's head around and started riding toward the inn in question.

'Is this a good idea?' asked Locklear.

'Probably not,' replied James.

As they approached the inn, a pair of moredhel warriors walked toward them. 'Where do you go, human?' asked one.

'Is that headquarters?' asked James.

'It is where Kroldech holds camp.'

'Is Shupik in there with him?'

'I know of no one named Shupik inside,' replied the guard.

'I guess he's not here yet,' said James, turning his horse off toward the centre of town.

They rode away and James said, 'Someone *really* doesn't like the idea of Kroldech being in command.'

'What are you thinking?' asked Locklear.

'Locky, my best friend, let's you, Patrus, and I go and see if we can sow a little dissent.'

Patrus chuckled his evil laugh as they approached another inn. Locklear and James dismounted, tied their horses to a line before the inn, and went inside with the old magician.

Pug sat wearily at his study table, in the small apartment set aside by Arutha for those times when he and Katala visited from Stardock. His eyes grew unfocused as he tried to read yet another report from one of Arutha's patrols, regarding an encounter with moredhel near Yabon.

He had spent hours sifting through reports, rumours and accounts from soldiers, spies and bystanders regarding the Six, Delekhan's mysterious magical advisors. The time he had spent with Owyn Belefote discussing his encounter with Nago, and what was before him now convinced Pug of an unsettling possibility.

He stood up and crossed to stand before a window that looked out over the harbour and the Bitter Sea beyond. Whitecaps danced on the sea as cold north winds cut down the coast. In the late-afternoon light, he could see ships racing for the harbour, attempting to reach safe haven before the storm arrived in full fury.

At times like these, Pug wished he had spent more time studying what was commonly known as the Lesser Path. Weather magic was an intrinsic part of that canon. His mind wrestled with a concept, one that he had been formulating for years, since he had returned to Midkemia as the first practitioner of the Great Path, as the Tsurani called their magic. Sometimes he felt as if he was peeling an onion, where every layer revealed only showed another layer below, made all the harder to perceive by the tears in his eyes. Then it hit Pug, *it's always an onion*.

He laughed. 'There is no magic. There are only onions!'

He knew he was too tired to continue, yet he returned

to the table. He had come to one frightening conclusion, a possibility he really didn't want to accept, but it was the only answer. Somewhere along the way, the moredhel had encountered and recruited a new ally.

A soft gong sound caused Pug to look up. The sound was a signal sent by a Tsurani Great One prior to arriving at the domicile of another, but he had not heard such a tone since leaving Kelewan, nine years earlier. He had no pattern here, so how his visitor had located him was a mystery.

The air before him shimmered for a brief instant, then Makala was standing before him. 'Greetings, Milamber,' said the Tsurani magician. 'Forgive the presumption of calling unannounced, but I felt it was time for us to come to an understanding.'

Pug said, 'How did you manage to arrive here without a pattern?'

Makala said, 'You are not the only member of the Assembly – '

'Former member,' said Pug. Despite the fact of his rank and powers being returned to him after the Riftwar, he had never returned to assume a position among the other members of the Assembly of Great Ones on the Island of Magicians on Kelewan.

'As you wish. *Former* member of the Assembly. You are not alone in your ability to progress beyond what many consider to be the conventional limits of our arts. I find that one can move at will to a location or person without the constraints of a pattern.'

'A useful ability,' said Pug. 'I would like to learn how to do it some day.'

'Perhaps some day you will,' said Makala. 'But I came here on another matter.'

Pug indicated a seat. The Tsurani magician declined. 'I will not be here long. I came to give you warning.'

Pug was silent. He waited and after a moment Makala continued. 'I and some of our brethren are involved in an undertaking that will not tolerate your interference, Milamber.'

'Pug,' he corrected. 'On this world I am Pug.'

'To me you will always be Milamber, the barbarian Great One who came to our world and sowed destruction among us.'

Pug sighed. He had thought that particular debate was a decade behind him. 'You're not here to revisit the past, Makala. What are you doing and what warning are you trying to convey?'

Makala said, 'What we are doing is of no concern to you, Milamber. And my warning is: do not attempt to involve yourself in any way.'

Pug was silent a long moment, then said, 'I know you were among those who were most resistant to my acceptance in the Assembly, all those years ago when Fumita brought me from the Shinzawai estate.'

'Resistant?' Makala smiled. 'I was among those who voted for your death before you entered training. I then considered you a grave risk to the Empire, and from my perspective, subsequent events bore out that suspicion.'

'Whatever I did, it was, in the end, for the good of the Empire.'

'Perhaps, but history teaches us that often such issues are merely a question of perspective. No matter. What is occurring now is being done without question for the good of the Empire, as is our mandate.'

Pug said, 'So then what I was on the verge of uncovering is now revealed to me by your appearance here.'

'What would that be?'

'That these magicians aiding Delekhan, the so-called "Six" are Tsurani Great Ones.'

'I congratulate you on arriving at that conclusion based upon evidence you didn't gather first hand. Impressive deduction, Milamber. But then Hochopepa always insisted you possessed an unusual mind.'

'It was easy enough if one paused but a moment to examine the behaviour of the participants in these various acts. The moredhel? They have always held a deep, abiding hatred of all other races, deeming anyone not of their people to be intruders in their domain. The trolls and goblins are often their tools.

'But when I looked at the pattern, I see gems from the Empire coming to Midkemia and being exchanged for gold. Had the gold returned to Tsuranuanni, there would have been no question, for there the gold is worth a hundred times more than here. But the gold never did. It went for weapons, and those weapons went to the moredhel. There was nothing in this for the Tsurani involved; nothing apparent.

'Then when reports of the magic used by Delekhan began to appear, things didn't fit. Some of the things reported could only have been done by Tsurani Great Ones.

'Which leaves me with this one question: why?'

'Why is not for you to know. Your judgment is called into question, Milamber. You revealed yourself as not being one of us when you destroyed the Emperor's celebration and drove the warlord to take his own life in shame. You live here, your birthworld, and you've taken a Thuril for your wife.

'You have a daughter who has shown power, yet you let her live.'

Pug's eyes narrowed in warning that his temper was about to come to play. 'Walk softly, Makala! This is not the Empire, and your words are not law.'

'We have difficulties on both sides of the rift,' said the Tsurani Great One. 'Others of our brethren now must deal with the consequences of the destruction of House Minwanabi by House Acoma. The order of the Empire is threatened. And here, on your birthworld, this academy you create at Stardock, why even some of our own have agreed to come teach your students.' His voice rose in anger. 'Our former enemies!'

'We are not your enemies,' said Pug, his fatigue suddenly threatening to overwhelm him. 'Ichindar knows this.'

'The Light of Heaven will not live for ever. Eventually, the Assembly will press for a return to the order we have enjoyed for two thousand years.

'But to ensure that you, the single biggest threat to our plans, do not interfere, we have arranged to take your daughter to a place where she will remain until such time we are satisfied you are no longer a threat.'

Pug's anger threatened to spill over. Barely able to hold back rage, Pug choked, 'Gamina! What have you done with her?'

'She is unharmed. She will remain safe as long as you do not attempt to hinder our plans.'

'Your plans involve murder on a wholesale scale if you're in league with the moredhel, Makala! Can you think I'd stand aside, even if it means my daughter's life, and let you destroy my homeland?' He moved to stand before the Tsurani Great One. 'And do you think to match your power with mine?'

'Never, Milamber. You are the greatest of our brethren, which is why you must be neutralized. But if you destroy me, there are others who will see that what must be done is done. We will not oppose you if you seek to reach your daughter.' He stepped aside and said, 'In fact, we will provide you with means to go to her, but I warn you this might prove a mistake, as even your daunting prowess will not prevail in returning you here.'

'Let me go to her,' Pug said, his fear for his daughter washing away his fatigue. 'As soon as I write a note to my wife.'

'No,' said Makala. 'If you go, you go now.' He took out a device, similar to a Tsurani transportation orb, but somehow different. He put it down. 'There is only one position, Milamber. It will take you to your daughter, but only if you leave within a minute of my activating it.' He clicked a slide on the side of it, and put it down on the pile of maps. 'That minute begins now.' He turned and walked away, producing another device and as he held his hand out to activate it, he said, 'My motives are for the good of the Empire, Milamber. I have never harboured any personal ill-will toward you. That is for lesser men. At the end of this, I hope you and your family are well, but if you oppose me, I will see you all destroyed, for the good of the Empire.' He vanished.

Pug grabbed a quill, dipped it in ink, and swept away all the papers and parchment on his desk, but one, a map upon the back of which he hastily penned six words. Then he dropped the quill and grabbed a writing charcoal, two pieces of parchment, and seized the device left by Makala, and with a fey humming, a high-pitched whine, the device activated, and he was gone, leaving only shifting papers on the floor as outside the window, the fury of the storm broke upon Krondor.

The inn was crowded, dirty and noisy, with men on the verge of brawling at the least excuse. James stood at the bar grinning.

'What are you so happy about?' whispered Locklear.

'I'm home, Locky. I've missed places like this.'

'You're crazy, boy,' said Patrus. 'You looking to die young?'

'I'll tell you about some of the places I spent my time in when I was a kid, some day. Right now I'm just enjoying the prospect of this bunch being the ones heading south in a few weeks.'

'Something's not right,' whispered Locklear. 'This isn't an army; it's rabble.'

'Locky, let's get some fresh air.'

He led his companions from the inn and outside. Evening had fallen, cold and damp, with a mist of rain starting to fall. When he saw they weren't overheard, James said,

'Everywhere I look I see wall fodder, with a few moredhel clans I would wager are not high on Delekhan's list of close friends.'

'Wall fodder,' chuckled Patrus. 'I like that.'

'Not if you'd ever had to be the first over the wall,' said Locklear who had stood on the walls at Armengar and Highcastle with James and watched warriors die trying to do just that.

'Where's the army?' asked James rhetorically.

'Moving toward us, even as we speak,' replied Locklear humorously.

'We might have a better idea if we knew what Kroldech knows.'

'Well, then,' suggested Patrus, 'why don't we just go ask him?'

James said, 'Or I could sneak in and see what he's got lying around that looks like orders.'

'You read that moredhel chicken scratching, boy?' asked the magician.

James lost his smile. 'No, I hadn't thought about that.' Orders from Delekhan to his field commander would be in that language, not the King's Tongue.

Patrus grinned. 'Well, I can.'

'How?' asked Locklear. 'Who taught you to read moredhel?'

'No one,' said the magician with a look of disgust on his face.

'Oh!' said Locklear, suddenly getting it. 'Magic!'

Rolling his eyes, Patrus said, 'Right, magic.' With a playful slap to the back of Locklear's head, he added, 'Idiot.'

James said, 'I think we have a problem, still.'

'What?' asked Patrus. 'You sneak in, get the papers, bring them out, I'll read them, you sneak in, put them back, and we leave.'

'That's the problem,' said James. 'I should be able to sneak in and out, once, but the second time? And if the plans are removed and found missing, they'll change them, almost certainly.'

'How many ways can they march down that pass and up to the walls of the keep?' asked Locklear.

'Several,' said James, 'and if we are ready for one, and they come a different way, well, even this rabble could create problems enough to cost us dearly.' He shook his head in frustration. 'Damn.'

They kept walking, not wishing to have anyone see them lingering. While most of the camp in the town was either asleep or drinking in one of the several taverns in Raglam, there were enough soldiers around to view with suspicion anyone loitering.

Locklear said, 'What if we could have a reason for being in there looking through the papers?'

'What?'

Locklear grinned. 'I have an idea.'

James said, 'I usually end up not liking it when you say that.'

'Come on,' said the younger squire. 'This is brilliant.'

'Oh, I *really* don't like it when you say that.'

Locklear crossed the largest street along the south end of town, and moved to the open field where the catapult aimed at town sat. A company of engineers lay sleeping at the base of the engine, and Locklear signalled for silence. He tiptoed to where the massive war engine sat and inspected it from a few feet away. Then he looked around on the ground until he found a rock the size of his fist. He pointed to the machine and in a whisper asked, 'Do you think you could hit that release lever from here?'

James looked a moment and said, 'No, but I could hit it from over there.' He pointed to a location the same distance from the catapult but at a different angle. 'I think you mean could I hit it and make it release?'

Showing frustration, Locklear said, 'Yes, that's what I

mean. Go stand over there, and when I signal you, count to one hundred. Then throw the rock and release the lever.'

'And what about the lashings?'

'I'll take care of that. Patrus, come with me.'

Locklear took the old magician and said, 'Walk around over there – ' he pointed to a location on the other side of where the engineers lay sleeping ' – and wait for me.'

Patrus headed off to do as he was bid, and when he saw James hadn't moved, Locklear shooed him away with a fluttering hand. James shook his head in disbelief, but he went where he was told to go.

Locklear crept close to the catapult and looked at the large restraining rope across the mighty engine's arm. If it wasn't in place, only the lever and gear arrangement kept the huge arm from discharging its deadly missile. As silently as possible, Locklear took his dagger and cut through the rope. It took several tense moments, as he sawed through the huge bundle of fibres, watching to see if any of the engineers stirred.

When the rope was severed, he moved away and quickly circled around the camp. He went to Patrus, took the old man by the arm and led him off into the dark. Just as he was about to vanish from sight, he signalled to James.

James, still not knowing what Locklear's plan was going to accomplish, counted to one hundred. When he reached seventy, he heard voices raised in the distance. When he reached ninety, he heard feet running in his direction. Not waiting to reach one hundred, at ninety-two he threw the rock. With his keen eye and strong arm, he put the stone right where it needed to be, knocking loose the lever. With a loud crash, the huge arm unloaded its stone, slamming hard against the crossbeam at the top of its arc. The sound instantly awoke the engineers who leaped to their feet, shouting. 'What was that? What? Who did that?'

Just then Patrus and Locklear arrived with a company

of moredhel warriors. 'There they are!' shouted Locklear. 'They tried to kill Kroldech!'

The warriors rushed forward while the still-stunned engineers milled around in mute astonishment. That lasted but a moment, then suddenly they were yelling at the moredhel guards, who were accusing them of treason.

Locklear took Patrus by the arm and hurried to James's side, while shouts and confusion came from the other side of the town.

'What did you tell them, Locky?'

'Just that this concerned old man, out looking for his lost cat, had come across this nest of traitors who were training their catapult on the commander's house, and he didn't know who to turn to, so I was bringing him over to that loyal bunch there.'

'Are they loyal?' asked James with a laugh.

Locklear returned the laugh. 'How do I know? Even if they're part of the faction trying to kill Kroldech, they're going to be all over those engineers for not waiting to do it when they were told.'

James spoke in appreciative tones. 'Damn, but you can be a sneaky bastard at times.'

'I take that as high praise, considering the source,' said Locklear.

They reached the area around Kroldech's headquarters and James said, 'I think I know what to do.'

He pushed through confused-looking soldiers and townspeople, saying, 'Stand away! Let us through.'

When he got to where he could see the damage he had to stop a moment in amazement. The stone had crashed through the centre of the roof, crushing the upper floor and collapsing it down on the second floor. The main doors were off their hinges. 'Damn, those guys were good,' whispered James in appreciation of the engineers' skill.

Then he realized he wasn't moving, and James said, 'We've got to save the commander!'

He waved at a few warriors nearby and said, 'Help us find the commander!'

They followed and James led them into the ruins of the inn. Several stunned warriors lay sprawled on the floor, and James had to duck under cracked and fallen ceiling beams, which were now only five feet above the floor in the commons. 'Where's the commander?' he asked one.

'He was over there, at his place in the rear of the commons,' said a moredhel warrior with blood running down his face.

Turning to those moredhel who had followed James inside, he said, 'Get these warriors outside to safety.' Pointing at Patrus and Locklear as if they were just two among many, he said, 'You and you, come with me and help me find the commander.'

They had to crawl under a beam. After a minute of negotiating their way in the gloom, they came to the room used by the commander. The door was off the hinges, and they had to climb over a fallen beam, but they got inside.

Two moredhel, killed by flying timber splinters the size of arrows, lay on the floor near the door. But behind a table crouched a moredhel, whimpering in terror, but otherwise uninjured. From the rings on his fingers and the golden amulet around his neck, James deduced he was the commander. He lay curled up and obviously shocked to near mindlessness.

'Not what one expects in a moredhel chieftain,' observed Locklear.

'Get him outside, Locky,' said James, 'but take your time, Patrus and I will see what we can save from the fire.'

'What fire?' asked Locklear.

James took paper and handed it to Patrus. 'Is this important?'

The magician closed his eyes a moment, then opened them. He looked at the document and said, 'No.' James took a shattered lantern, and dipped the paper in it. Then he produced a flint and steel from his belt pouch and struck sparks on the paper. It ignited. Taking the burning paper from Patrus, he pointed to it with his other hand and said, 'That fire.'

Locklear grinned. 'Oh.' He pulled on Kroldech's arm and said, 'Commander, we must flee! Fire!'

That seemed to energize the stunned moredhel chieftain. He let Locklear help him to his feet and said something in his native tongue.

'Come with me, Commander,' Locklear repeated. He led Kroldech away.

Patrus and James quickly examined papers, and each one that Patrus gave James that wasn't important, James added to the growing fire.

Finally, he said, 'This. This is the attack plan.'

'Read it to me,' said James, 'quickly.'

Patrus did and James forced himself to remember every word as it was being read. 'I have it. Now, grab up some other papers and follow me.'

The fire was now burning in earnest, and by the time they reached the point where they had to crawl under the timbers, it was getting hot. Just as flames erupted through the roof they reached safety outside and found Locklear holding up the still-wobbly commander.

Reaching them, James said, 'Master! We managed to save these papers.' He held out the entire random bundle of papers.

Kroldech's eyes focused and at last he understood what happened. 'Assassins!' he shouted. 'They tried to kill me.'

'They are in custody,' said the moredhel chieftain, who had been alerted by Locklear. 'These mercenaries saved you, master.'

Kroldech grabbed the papers from James and started inspecting them. After a moment, he came to the orders of battle, and smiled. 'Good!' He struck James on the arm, hard enough to hurt. 'You are heroes!' He stuck the battle plan under James's nose. 'Do you know what this is?'

James feigned ignorance. 'No. We just grabbed what we could, master.'

'If this had been lost, I would have had to redraw all our plans. You've saved me days of labour.' Looking at the fire, he said, 'And you saved my life. I am in your debt.'

'Think nothing of it,' said James.

'Nonsense,' said Kroldech. 'Come to me tomorrow and I will reward you.'

'Thank you, master,' said James. 'We will.'

The still-shaken moredhel leader allowed himself to be escorted away to new quarters as James turned to Locklear and said, 'Where's Patrus?'

'He was with you. Maybe he's over where our horses are waiting?'

They walked to where their horses were waiting. Patrus had a third horse and was mounting it. Locklear said, 'Kroldech said we're heroes. Wants us to come by tomorrow and collect a reward.'

'You going to hang around for the reward, James?' asked the old magician.

'When trolls can fly. By tomorrow morning, I want to be halfway to Northwarden.'

As all eyes were on the burning inn, they slipped out of town, and managed to get down the road before being challenged. The bored-looking mercenary asked what they were doing on the road late at night and James said, 'The elves can't handle those trolls down south, so we're being sent to sort them out.'

'Heard there was some trouble down there,' said the guard. 'Good luck.'

'Thanks,' said James.

After they were out of earshot, Locklear said, 'Patrus! Where did you get that horse?'

'I borrowed it,' said the old magician with a cackle. 'Kroldech won't miss it until tomorrow.'

Locklear's only satisfaction on the way back was that James had to spend his pouch of gold to get past the trolls, but at least the trolls thought of them as friends now. The ride was difficult, as the weather had turned very cold and wet. The horses were tiring, and had to be walked at times.

Eventually they reached the road up to the keep and James said, 'Where are our soldiers?'

Locklear said, 'I thought some of the forward elements might be trying to keep out of the rain, but you're right. We should have seen others by now.'

James set his heels hard against his horse's sides and was off at a canter, demanding as much as the fatigued animal could give going up the steep road to the keep. When they were within sight of the keep, they saw the gate was up and the portcullis down, and torches burned on the walls.

'They've crawled inside and buttoned up!' said Locklear.

Reaching the edge of the moat, James called out, 'Hello the castle!'

From above a sentry shouted, 'Who goes there?'

'Squire James, Squire Locklear and Patrus. Let us in.'

There was some discussion, but eventually the massive bridge was lowered while the iron lattice of the portcullis was raised. James and the others rode across the drawbridge.

Inside the barbican, a group of soldiers waited, and James dismounted. 'What is wrong?' he demanded.

A soldier said, 'Assassins, squire. Nighthawks in the castle.'

Locklear said, 'What has happened?'

'Baron Gabot is dead, squire. Two captains, and our sergeant.'

'Gods,' said Locklear.

'Who's in charge?' asked James.

The soldiers exchanged glances, and finally one said, 'You are, squire.'

FOURTEEN

Instructions

Riders hurried along the highway.

Owyn, Gorath and Ethan Graves rode quickly down the highway toward Krondor. They had spent one night at Darkmoor, in a decent inn, indulging themselves in a bottle of good wine – which Gorath grudgingly admitted was better than that served by Baron Cavell – and a hot meal before sleeping on down-stuffed mattresses. The rest of the journey had been less hospitable, sleeping under the stars away from the road, bundled up in sleeping cloaks on rocky ground, and only twice in the rain.

They had made good time from Malac's Cross to Krondor – less than fifteen days – and hadn't killed their horses in the process. Now they were within sight of Krondor.

As they slowed their horses to a walk, Graves said, 'I must throw myself on the mercy of the Temple of Ishap and confess my sins.'

Owyn said, 'What will they do?'

'Execute me, perhaps, or exile me. I don't know.' He sighed. 'I don't much care, but before that I have to get Kat out of the city.'

'Where will you send her?'

'To Kesh. I have connections there. Old trading partners in Durbin.'

Owyn said, 'From what I hear Durbin's a rough place.'

'So is Krondor if you have to live on the street,' said Graves.

Owyn was still trying to piece together all the relationships he and his companions had uncovered since he had first met Locklear. He wished more than once that Squire James was still with them. He asked Graves, 'What about the Prince's justice?'

Graves shrugged. 'If the Ishapians turn me over to Arutha, he'll probably hang me.'

Owyn reflected on that. In the two weeks he had spent in Graves's company he had come to like the gruff old man. He was unapologetic about his early past, simply admitting he had been involved in smuggling, extortion, and had killed more than one man on behalf of the Mockers of Krondor. He made no brief excusing his behaviour and only said that since he had heard the call of the temple, he was a changed man.

Owyn believed him, but also decided if a fight broke out he'd want Graves on his side. He was still a powerful-looking man despite his grey hair and lined features.

The gate to the city was manned by armed guards, one of whom put up his hand and said, 'Halt!'

Owyn said, 'Trouble, guardsman?'

Pointing at Gorath, the guard said, 'Who's this?'

'You can talk to me,' said Gorath. 'I speak your language.'

'Well, then, who are you?' demanded the guard. 'What's your business in Krondor?'

Gorath said, 'I bring a message from Prince Arutha to the magician Pug.'

The guard blinked in surprise at the mention of those names. He motioned them aside and said, 'We'll have you escorted to the palace.' His tone made it clear this wasn't optional. Another soldier hurried into the city and returned less than ten minutes later with half a dozen burly men wearing the tabards of the city constabulary. At their head was a tall man who bore a badge of office on his tabard. He conferred a moment with the sergeant at the gate then came to stand before Gorath. 'You claim to be carrying a message from the Prince to the magician Pug?'

Gorath replied, 'That is what I said.'

'I am the Sheriff of Krondor. Is there someone at the palace who can vouch for you?'

Gorath glanced at Owyn. Owyn said, 'We met a lot of people, but most of them are out in the field with Prince Arutha. If Pug is at the palace, he'll vouch for us.'

The sheriff spent a moment casting a baleful eye on the three of them, then said, 'Come along.'

He started toward the palace, and Graves said, 'I have to get to the Temple of Ishap.'

Over his shoulder, the sheriff said, 'You can visit the temple after I leave you at the palace. We've got orders concerning the comings and goings of suspicious-looking individuals, and you fit the description. If the Captain of the Royal Guard turns you loose, that's his decision.'

'I am a member of the Order of Ishap, and I am under their protection,' said Graves.

'Then they can come and fetch you out if the captain has any problems with your story,' said the sheriff in a no-nonsense tone.

They reached the palace without any further conversation, and at the gate the sheriff turned them over to the Royal Guard. A sergeant came and said, 'You lot look familiar enough, but I've no orders, so let me send word inside about what to do with you.'

Again they waited, and after a while a message came telling the sergeant to admit the three men. The sergeant ordered palace grooms to come take the horses and palace porters to carry their bundles inside. Then he led the three of them to the office of the Knight-Marshal.

A captain sat alone and looked up when they entered. Owyn didn't know his name, but he had been present when last they had spoken with the Prince, and would know they were who they claimed to be. 'Owyn,' he said in greeting. 'You have a message for the magician Pug?'

'Yes,' said Owyn. 'From Prince Arutha. He wishes the magician to join him, as he fears magic will come into play soon in the coming invasion.'

The captain, a veteran of long years of service, looked openly frustrated. 'I would prefer nothing more than to oblige my liege lord, but at present, the magician Pug is absent.'

'Has he returned to Stardock?' asked Owyn.

The captain shook his head. 'No one knows where he has gone. His wife came to us a few days ago with the news he had vanished in the night, leaving only a cryptic note. More than this, no one knows.'

Gorath said, 'Could he have been abducted?'

The captain shook his head again. 'I know little of magic, but my understanding of Duke Pug's talents leads me to believe had he not left of his own will, much of this palace would be smoking rubble.'

'May we see this note?' asked Owyn.

'You'll have to take that up with the Lady Katala. I'll send word and see if she wishes to speak with you.'

A page returned quickly with word the Lady Katala indeed wished to speak with them. They hurried after the page to the private apartment set aside for Pug and his family when visiting the palace, and found Katala waiting.

She was a striking woman, despite her diminutive size, dark complected and showing a slight dusting of grey in her otherwise dark hair. While small, there was a strength about her that made her distress all the more apparent. She was close to being frantic, yet her emotions were under control.

Her accent was strange to Owyn, something akin to that of Sumani and the other Tsurani he had met in Yabon, but not quite the same. She said, 'I understand you come seeking my husband?'

'Yes, lady,' said Gorath. 'We carry word from the Prince that Pug is needed.'

'Where is he?' asked Owyn.

'I don't know.' She paused. 'You remember our daughter, of course.'

Gorath nodded.

'She went missing a few days ago, and I went seeking my husband in his tower. He also was missing.'

'Perhaps they went someplace together,' suggested Graves.

Katala looked at the stranger and asked, 'Have we met?'

Owyn introduced them, and Katala said, 'Abbot, my husband would never have left this message had that been the case.'

She held out a parchment, upon which was written, 'To Tomas! The Book of Macros!'

'What does this mean?' asked Owyn.

'Tomas is Pug's childhood friend,' said Katala. 'He is now living in Elvandar.'

Gorath asked, 'The wearer of the white and gold?'

Katala said, 'Those are his colours.'

Gorath said, 'There have been stories among my people, that when those who travel from the Lake of the Sky to the Green Heart come too close to the borders of the land of the eledhel, occasionally one garbed in the raiment of the Valheru will appear. His powers are terrible.'

'Those are not stories,' said Katala. 'Tomas exists, and he may be the only one on Midkemia with enough power to find my husband and daughter.'

'Did you send anyone to carry word to him?' asked Owyn.

'Not yet. The Prince took most of the army with him. Those left in charge, like the captain of the Royal Guard and the Sheriff of Krondor, are unwilling to exercise discretion beyond what they see as the clear requirements of their offices. Most of the other nobles are with the Prince or upon other business here in the west.' She looked very distressed. 'There really isn't anyone to send, and I'm not even sure if this message is intended for Tomas.'

Gorath said, 'Perhaps Pug is instructing someone to take this Book of Macros to Tomas?'

Katala said, 'I helped my husband catalogue the entire collection Macros left behind at Sorcerer's Isle, including those left behind there and those sent to Stardock. There was no single volume I'm aware of called "The Book of Macros", so it may mean something else.'

Owyn looked at Graves and Gorath. 'Perhaps we should take this parchment to Elvandar?'

Graves said, 'As much as I am in debt to you, Owyn, and your friends, my life is held by a short thread. I must make my way to the Temple of Ishap and face my punishment.' He glanced around, as if fearful of being overheard. 'If those here who have authority know a tenth of what I have done, I would be in the dungeon below, I am certain.'

Katala looked confused. 'Perhaps we can help?'

Owyn held up his hand. 'Lady, he speaks true. He was moved by his love for another, but he has betrayed his nation and his temple.'

Graves said, 'I must go to the temple and make my confession. If you will excuse me, I will leave.' Taking Owyn by the elbow, he led him aside and said, 'On your way north, stop at the Abbey of Sarth. They will have knowledge of this Book of Macros if anyone other than Pug does. Besides, they should know of what we have seen.'

Owyn said, 'I was hoping to take ship.' He glanced at Katala. 'If the magician's wife can arrange it.'

'Take ship from Sarth,' said Graves.

Owyn looked as if he had no better suggestion. 'Very well. What will happen to you?'

Graves shrugged. 'Expulsion, certainly, and shame. I may be given a chance to redeem myself through years of penance, but I think I will be put on the street and told to leave. Perhaps a grace period or they'll alert the Crown I have committed treason and the watch will be waiting for me when I leave the temple.' He seemed fairly indifferent to his own fate, but his manner and voice changed when he said, 'But I must get Kat out of the city and safely away. I did this only to protect her, and if I fail in that, all is futility.'

'How will she get away?'

Graves smiled. 'My Kat is a woman of no mean talent and wiles. She has her route out of the city already chosen,

I imagine, and if I send her word, she will be gone by morning.'

'Can you get her word?'

'If I can reach someone in the Mockers, no doubt.'

'Then fare you well, Abbot.'

'Fare you well, Owyn.' He turned to face Gorath. 'Take care of yourself, as well.' He bowed to Katala. 'Lady, good-bye.'

He left.

Owyn turned to Katala and said, 'Lady, if you can facilitate getting us the means, we will take this note to Elvandar.'

'What do you need?' she asked.

'Funds, I fear, for we lost most of ours in the north. Fresh horses, so we may ride to Sarth. Then we should sail to Ylith and take horse to Elvandar. I fear I am asking a great deal, and you know little of us.'

'I know that my daughter touched Gorath's mind, and after she said she felt no malice in him toward us.' She looked at the dark elf and said, 'I find it odd, for all I have ever heard of your race is an abiding hatred of ours.'

Gorath said, 'Two years ago, Lady, I would have found it equally odd. All I can say is that life has turned and things are not as they once were.' He stared out a window that overlooked the city. 'The world is much larger than I once dreamed, or perhaps my place in it is smaller than I once realized.' He shrugged as if the difference was unimportant. 'But whichever is true, it is far more complicated a place than I had ever imagined in my years in the icy north.' He went to the window and gripped the edge of it, his voice dropping. 'I will help because I once had children. I can't say more of them, for the pain still lingers, and that wound will not heal.' He looked at Katala. 'I will help find your husband, and I will bring your child home to you.'

Katala, born of a race of proud warriors, looked at the moredhel chieftain and her eyes were bright. No tears fell, but it was clear to Owyn that Gorath's words had reached her. 'I will see what I can do,' she said softly. 'Wait here.'

She left and Gorath and Owyn sat. Owyn said, 'Is it safe for you to travel to Elvandar?'

Gorath smiled at Owyn and said, 'We'll find out, won't we?'

A note from Graves arrived at the palace the next day. It said, 'We're fleeing to Durbin. Tell Jimmy I'm sorry. Graves.'

With the note was a hastily drawn map, and some instructions on how to operate a secret entrance to the abbey from an abandoned dwarven mine below. Scrawled at the bottom was the note, 'In case you have trouble getting in.'

Katala arranged for horses and enough gold to secure passage by ship from Sarth and get more horses when they reached Ylith. The captain arranged for them to be accompanied by a patrol as far as the road leading to the Abbey of Sarth and they left the following day.

Owyn memorized the map Graves had sent, then asked Katala to see that James got it if he returned.

The trip north was uneventful, either because things had moved to a point where Gorath's freedom was no longer important to Delekhan and his agents, or because they just didn't know where he was any longer. At the base of the road to the abbey, they parted company with the Kingdom soldiers and headed up the hill.

As they rode, Gorath said, 'This must have been a fortress once.'

'I believe it was, a robber baron's or something like that. The Prince of Krondor at the time gave over the property to the Ishapians.'

As the road rounded a curve, Gorath said, 'It must have been a murderous battle to storm that position and take it.' He pointed to the abbey, now visible at the top of the mountain. High walls close to the sides of the cliffs provided a daunting image. Owyn was forced to agree that he would not wish to be among those storming this old fortress.

They reached the gate and Owyn shouted, 'Hello, the abbey!'

To the right of the gate a figure appeared up on the wall. 'Hello, travellers. What do you seek at the Abbey of Ishap at Sarth?'

'I carry a message from Abbot Graves, late of Malac's Cross.'

The figure disappeared and a moment later, the large door swung open. As they rode in, it swung closed behind them and a very old monk, carrying a large warhammer, stood behind them. 'By the beard of Tith! A Dark Brother riding into the abbey like he belonged here.'

Another monk put up his hands in a calming gesture. 'Brother Michael, these fellows say they carry word from Abbot Graves at Malac's Cross.' He turned toward the two as they dismounted and said, 'Brother Michael is our Keeper of the Gate. Earlier in his life he was a warrior, and occasionally he falls back into the habits of his youth.'

Gorath studied the grey-haired old man, still upright and strong, despite his age. With a slight incline of his head, he showed his respect. 'If his task is to be vigilant, he serves you well,' said Gorath.

'I'm Dominic, Brother Prior to the abbey and in the Abbot's absence, I am in charge. What may I do for you?'

Owyn introduced himself and Gorath and replied, 'We travelled with Squire James of Krondor, and made the acquaintance of Abbot Graves on our way to Romney a few months ago, and we had reason to visit with Graves recently. He travelled with us to Krondor, to throw himself on the mercy of the temple.'

'Come inside,' said Dominic. He motioned to a monk to take their horses. 'Please, follow me.'

Dominic appeared to be a middle-aged man, but one who moved with a quick step. His dark hair was showing grey, yet there was a light of curiosity in his eyes that was refreshing. He showed them to an office and said, 'Please, sit down. Would you care for something to drink?'

'Water, please,' said Gorath.

Dominic asked a monk to fetch mugs of water and said, 'I remember James from a visit here many years ago. He was quite a personality.'

'He still is,' said Gorath.

Owyn smiled at that. 'Abbot Graves asked me to tell you what has occurred.' He summed up what he knew, then filled in details when Dominic asked him some questions.

Finally, Dominic observed, 'Well, this is a matter for the mother temple in Krondor, but I fear the Abbot will be subject to the most severe punishment.'

'Why?' asked Gorath.

Dominic looked at the dark elf. 'Why? For betraying us, of course.'

'Do you fault the tool for bad work, or the worker?'

'I don't take your meaning,' said the monk.

'Your order selected this man. You subjected him to whatever rites and oaths you human priests use. Yet you admitted a flawed man to your ranks.'

Dominic sighed. 'We are not perfect. We make mistakes. It was a mistake to admit Ethan Graves to our ranks, no matter how urgent he felt his calling was.'

Owyn said, 'Well, at least he returned to pay his debt.'

Dominic sat back. 'I wonder . . .' After a moment of reflection, he stood up. 'In any event, I cannot help you in the matter of this Book of Macros you mentioned. Pug allowed us to copy certain volumes in his library in exchange for our sending him copies of a few volumes here in our library.'

'Could the Book of Macros be something that's stored here, without your knowing?' asked Gorath.

Dominic motioned for them to walk with him. 'No, every volume in our possession is catalogued and can be found easily by our master librarian.' He took them through the main building of the abbey, and said, 'Rest, and eat with us. I will send one of the brothers into town to inquire after the next ship bound for Ylith. If you leave your horses with us, you may reclaim them should you come this way again.'

'Thank you,' said Owyn. They were shown to a room with two narrow beds. Gorath lay down and was quickly asleep. Owyn lay down, but sleep was slow in coming as his mind wrestled with questions for which he had no answer: what would happen to Graves and his Kat? Where were James and Locklear? And most of all, what was the Book of Macros and where could they find it?

James looked at the maps and shook his head. 'We just don't have enough men.'

His makeshift staff stood arrayed around the table. James had appointed new commanders, based on quick interviews with various soldiers in the keep. He had appointed temporary officers and sergeants, and reorganized patrols and duty rosters. The past week had seen things start to firm up, but now he was getting reports of troop movements to the north.

'Whatever trouble we caused with our pranks up there seems to have finally been overcome,' he said to Locklear, who stood to one side. 'It's clear they're starting to stage for the move south. Another month at the outside and they're going to be heading our way.'

'Should we try to send another messenger south?'

'The Earl of Dolth has an outpost on the northern edge of the Blackwood. That's about the only place we haven't sent a messenger.' He looked around the room. 'No, we're here, and unless help is already on the way, we're on our own. See to your posts and try to keep a brave face; our men need it.'

Locklear said, 'Should I ride out and take another look?'

James shook his head. 'No, they're coming. This report says there are siege towers on the move, as well as catapults.'

'Then what next?'

'We wait,' said James. 'Have a patrol sweep south and west, to make sure we don't face surprises from unexpected quarters, then have the word go out to the surrounding villages.' James had recalled the horse soldiers from the town of Dencamp-on-the-Teeth, and was using them for patrols. That also had gained him one sergeant with experience. 'I want militia gathered and brought here and those who won't or can't fight sent to the south.' He pointed to the map. 'Start digging traps here, in the morning. By the time they get here, I want their engineers having to fill pits all the way up that road.'

Locklear nodded. 'Shall I have crews start bringing up boulders?'

'Yes. There's a ridge here – ' his finger touched a spot on the map ' – where a ledge overhangs a curve in the road. If you build a wooden cradle and fill it with boulders, we can pull out supports and rain stones down on them.' He considered his situation and said, 'If they don't bring magicians against us, we might possibly keep those damn siege towers away from our walls.'

'Bah!' said a voice from the corner, and James and Locklear turned to see Patrus standing a short distance away. 'If they bring their spell-casters, I'll show them a thing or two.'

James smiled. 'Good. We'll rely on you.' He looked at his long-time friend. 'Any luck in finding the assassins?'

Locklear shook his head. 'And I'm worried. It could be someone in the garrison, in the staff, or someone who snuck in and then left. I don't know. Two of the captains were killed while in the field, sleeping in their own tents, and the Baron was poisoned, while no one else at the table suffered so much as heartburn.'

'So we may have several Nighthawks still among us?'

'Yes,' said Locklear. 'I wish we had a way to ferret them out.'

'Let me roast a couple of prisoners over a fire,' said Patrus with an evil cackle. 'That'll scare the rest of them into confessing.'

James paused a moment, and Locklear said, 'You're not thinking of taking his suggestion seriously, are you?'

'No,' said James with an impatient shake of his head. Then suddenly his grin returned, and he said, 'But it gives me an idea.' Turning to Patrus, he said, 'Can you keep a secret?'

'Of course not,' said the old man, then he laughed at his own joke.

'Good, because I have a secret I want you to keep, for, oh, a few minutes at least.'

'What's that?' asked the old magician with as delightedly evil an expression as Locklear and James had ever seen.

James began to outline his plan and the magician began to chuckle again.

James and Locklear stood above the common dining hall, looking down from a balcony that led into the late Baron's meeting room. Soldiers were talking over their meal, their voices low. Locklear said, 'It's spreading.'

'Like a rash,' said James.

'When do you think they'll act?' asked Locklear.

'If I know my Nighthawks, the second they think there may be a way to discover who they are, they'll be looking for a way out. The longer they wait around, the higher the chance of being discovered.'

'You think they'll believe Patrus?'

'Why wouldn't they?' James asked. 'Most of these soldiers know nothing about magic. As a group they're tough, good fighters, and not very bright; else they wouldn't be up here on the border.'

'I can't argue that,' said Locklear, who had spent more time on the border than James. 'You usually have to be pretty stupid to get banished up here.'

'Or if you volunteer, you're even dumber,' offered James.

'Anyone looking nervous?' asked Locklear.

'Over there, those three in the corner.'

Locklear watched as three soldiers who had been sitting by themselves, huddled with heads low over the table, talking among themselves and trying hard not to be overheard. One seemed to be arguing with the other two, but whatever they were discussing, they were keeping it to themselves.

Finally, the other two seemed to convince the dissenter of what it was they had been arguing, and the three of them stood up, one of them looking around the room suspiciously, while the other two tried to appear casual.

'Is the gate closed?' James asked.

'Of course, as you ordered.'

'Then it's the postern gate,' said James.

'What about the sally port?'

'Too close to the front gate. No, they'll try to sneak out the back way. Besides,' said James with a smile, 'I left it intentionally unguarded. An "oversight" by an "inexperienced commander".'

'You're an evil bastard, Jimmy the Hand.'

'Why, thank you, Locky!' said James brightly.

They moved away from the balcony and hurried down a flight of stairs, where two men they had decided could be trusted waited. The old sergeant said, 'I saw three men leave a moment ago, squire.'

'Do you know them well?'

'No. Two of them came in last summer, replacements from Romney, and the other came here but a few weeks ago.'

James nodded. 'They're the ones. If we check with the other men in the command, I'm willing to wager one of them was working in the kitchen the night the Baron died, while the other two were with the two dead captains.'

'Where are the others?' asked Locklear.

'There are ten men I know I can trust, squire,' answered the sergeant. 'Most have been here for years, and one is my brother's son. They're all waiting near the stable.'

'Good,' said James. 'Let's go.'

The four of them hurried through a tunnel at the rear of the keep, and came through a door that opened into the stabling yard. As James had anticipated, the three suspected Nighthawks were hurrying toward the stable.

The old sergeant put fingers between his teeth and whistled shrilly. From the stable ten soldiers appeared, running at the three Nighthawks.

Instantly one of them turned and saw the four coming from the rear. Seeing that they were surrounded, they offered no resistance. But as James neared, he saw all three put their hands to their mouths. 'Stop them! They're swallowing poison!'

Soldiers sprang forward, but it was clearly too late. By the time they reached the three the Nighthawks were already falling to the ground, their eyes rolling back into their heads, and their bodies twitching uncontrollably.

'Damn fanatics!' said James.

'Who are they, squire?' asked the old sergeant.

'True Nighthawks. Perhaps some left from the Great Uprising or others recruited since then, but willing to kill and die for dark powers.'

He looked at Locklear who nodded. 'Search them for any papers then burn the bodies,' said Locklear. 'Now.'

'No priest?' asked the sergeant. 'There's a temple shrine to Lims-Kragma down in the village of Putney.'

'No,' said James. 'Burn them within the hour. I want to make sure they stay dead.'

'Stay dead?' asked the sergeant.

James didn't answer. No sense alarming the men, but he all too vividly remembered those Nighthawks in the basement of a brothel in Krondor who rose to kill only minutes after dying themselves. He hoped he would never see anything like that again.

'What do we do now?' asked Locklear as he overtook his friend.

James said, 'Sharpen our swords, oil our armour, and wait for Arutha.'

Owyn had never liked sea travel, and Gorath admitted it was an alien experience to him. Yet both managed to bear up under the swift voyage from Sarth to Ylith. Favourable winds and no encounters with marauding Quegan war galleys had kept the journey to under four days.

At Ylith they had purchased horses with the gold given them by Lady Katala and after consulting with the local garrison commander, discovered that things had turned quiet in the west. Whatever attempts Delekhan had made to convince the Kingdom he was attacking in the west had failed and the attempts had been abandoned. Owyn could only conclude that was because the enemy now were preparing to direct their attentions elsewhere.

Gorath pointed and said, 'On the other side of those mountains lies the Green Heart. There hide some of my people opposed to Delekhan. They will aid us if we find them.'

'According to the captain in Ylith,' replied Owyn, 'we should find ourselves in dwarven territory, near a place called Caldara. The dwarves should be willing to help us get to Elvandar.' Gorath's expression clearly showed he thought that an unlikely turn of events.

They rode toward Zūn, where they would take a road into the mountains, which should be clearing of snow as spring approached. The garrison commander had given them warning that the short route to Elvandar was the most dangerous and that if they wanted a safer way, they should go north to Yabon, then westward along the River Crydee from the Lake of the Sky, but that would add a month's travel. Owyn and Gorath were both feeling that time was now their enemy.

The attack would come soon, for any timetable that sought to put an army in Sethanon by summer would have to begin soon. No matter which route Delekhan's forces took, they would have hundreds of miles to cover, and supplies would be a problem. Forage along the way would be best in spring and summer.

Owyn knew that even as they rode the enemy might be launching his invasion of the Kingdom.

'Where are they?' demanded James. He stood on the battlements of Northwarden, staring up the gap as if he could see into the Northlands. He had expected the attack a week earlier, and still there was no sight of the enemy.

'Should I ride up and take another look around?' asked Locklear.

'No. It will probably look the same as the last time, lots of warriors gathering and arming.' James tried not to let the frustration show, but it was difficult. 'They will come when they do, and there's little we can do but wait.'

'At least Arutha and the relief should be getting here sooner,' said Locklear.

'Yes,' said James, 'if Owyn and Gorath got through.' Then he looked down the road toward the enemy. 'But if they had, I would have expected Arutha to be here by now. Something must have happened to them.'

'Then you think we're not going to get help?' asked Locklear.

James shook his head. 'There's no force of size in the east close enough to help. Other than the border barons, all our forces are in the south, near the Keshian border or in the east, ready to deal with the eastern kingdoms.'

Locklear sighed. He looked at James, then he smiled. 'Well, it's not the first time we've found ourselves in a hopeless situation, is it?'

James said, 'No, but it's the first time we've been in charge of a hopeless situation.'

Locklear's smile faded.

FIFTEEN

Quest

Winds cut through the pass.

Gorath and Owyn pulled their cloaks tightly around them as they rode. It was spring, but the mountains still held firmly to winter.

Gorath said, 'We're being watched.'

'Who?'

'I don't know. But I've seen movement along the ridge above us for the last hour. If they meant us ill, they would have attacked by now.'

A few minutes later, a figure wrapped in a heavy cloak appeared on a rock ahead of them. He stood waiting.

As they drew closer, Owyn saw it to be a single dwarf. He held up his hand in greeting. Gorath reined in and said, 'Talk to him first, Owyn.'

Owyn nodded and moved ahead of Gorath, letting the moredhel follow a few paces behind. When they reached a point near the dwarf, Owyn stopped, threw back his hood and said, 'Hello.'

The dwarf threw back his own hood, revealing a black beard of awe-inspiring thickness and hair that refused to be organized into anything remotely coherent; the moustache stood out like a huge bristle-brush. The dwarf's eyes went from one rider to the other as he regarded both with suspicion.

'Greetings,' he said calmly. 'What brings you two up into the frosty passes of the Grey Towers?'

Owyn said, 'We carry a message from Lady Katala, wife of Pug the magician, to Tomas, Warleader of Elvandar.'

The dwarf scratched his chin. 'That's a good one. I've not heard it before. In fact, I'm inclined to believe you.'

Owyn said, 'Why wouldn't you?'

The dwarf pointed at Gorath. 'His kin have been coming down from the north for the last year or better, and we'd forgotten how irritating they could be as neighbours.'

Gorath pulled back his hood, and said, 'I doubt they feel any more warmly toward your people, dwarf, but the problems between your people and mine are for another time. Right now we need safe passage to Elvandar.'

The dwarf squatted atop the rock and said, 'Elvandar? Well, if you say so. As I understand such things, you're likely to get even less warm a welcome from your cousins up there than you will from my folks.' Looking at Owyn he added, 'You wouldn't have any sort of warrant from someone in authority now, would you?'

Gorath nearly spat with contempt. 'And what gives you the right to ask for such a thing, dwarf?'

'Well, to begin with, you're on my land. Then there's the twenty of my people who have surrounded you while we talked.' He whistled, and, seemingly out of nowhere over a score of dwarves stood up. Owyn saw they all were heavily armed.

'Point well taken,' said Owyn. He reached into his tunic and pulled out a message from Katala, bearing a ducal imprint and a countersignature from the Captain of the Royal Krondorian Guard.

The dwarf glanced at it and handed it back. Then with a grin he said, 'I believed you from the first. Say what you will about the moredhel, they've never been demonstrably stupid, and riding in here in plain sight would be exactly that if you were planning mischief. Come along, we'll escort you into the village.'

'Village?' asked Owyn. 'Are we near Caldara?'

'Another half an hour. You can explain what it is that's got you in such a hurry to reach Elvandar.'

'Explain to whom?' asked Gorath.

'King Dolgan,' said the dwarf. 'Who else?'

Nothing more was said as they moved along the trail, and when the cut-off appeared they followed it down into a small valley, in which nestled a pretty little village. All the buildings were whitewashed stone with thatched roofs, save a large wooden hall with a heavy log roof which dominated the centre of the village. They made for that building, and the dwarf who had led them said, 'The lads will take care of your horses. The King is inside the long hall.'

They were at the narrow end of the long hall. Owyn and Gorath mounted stone steps into the building. As they reached the door, the dwarf halted. 'Present yourself to the King. I will see you later.'

Owyn said, 'Are you coming in?'

The dwarf shook his head. 'No, I have other business. You'll be able to find your way. Just follow the passage to the end of the corridor and you'll see the King.'

Gorath said, 'You've been hospitable, dwarf. I would know your name.'

The dwarf smiled. 'I am Udell. I am the King's younger son.'

Owyn opened the door and found himself looking down a long hallway with doors on either side, at the far end of which he could see a large room. He moved down the corridor, and when he and Gorath reached the end of the hall, they entered a common room dominated by a large square formed by four long tables. At the closest corner sat five dwarves. One of them stood and announced, 'I am Dolgan.'

Owyn bowed awkwardly and replied, 'Your Majesty.'

Dolgan waved away the title and said, 'Just Dolgan.' He tamped down a pipe and lit it with a smouldering taper. 'Now, what brings you two to Caldara?'

Owyn said, 'Lady Katala, wife of Pug the magician, asked us to carry a message to Warleader Tomas in Elvandar.'

Dolgan raised an eyebrow. 'Tomas is an old and dear friend.' With a smile he added, 'An uncommon lad.' He glanced at Gorath and observed, 'You pick unusual companions, boy.'

Owyn said, 'Gorath brought warning to the Prince that a leader named Delekhan was mounting an invasion.' He went on to explain the entire situation to the dwarven king, who listened without interrupting.

After Owyn was done, the old dwarf sat silently for a while, weighing what he had heard. Then he looked at Gorath. 'Well, my old enemy, answer me one question: why do you warn your enemies so that we may slaughter your kin?'

Gorath was silent for a moment as he considered his reply; then he said, 'I do not wish to see my kin die. I wish to see Delekhan overthrown. It has gone too far, and too few of us oppose him, but should the Kingdom defeat him, Delekhan will lose his hold upon my nation. Then many of us will rise up and depose him.'

'Then what?' asked Dolgan. 'Another warlord to rally around? Will you take his place?'

Gorath looked at the old King and said, 'I think I will never again see the Northlands. Two wives, two sons and a daughter have I lost. All who are blood kin are dead. I have nothing there. But whatever may occur in the future, well, I cannot speak to that; I can only say that Delekhan must be stopped.'

Dolgan nodded once, emphatically. 'Well said. We shall help you. During the Riftwar my people would move to Elvandar to fight with Tomas and the elves every year. We have a safe route that will take you close to their border and from there you can make your way safely to the Queen's court. I'll send along a few of the lads to ensure those of your kin and some goblins who've been pestering us lately don't give you any trouble.' He stood up. 'Now, rest and eat and tomorrow we'll have you on your way.'

Owyn said, 'Thank you . . . Dolgan.'

The dwarven king smiled and said, 'That's it!'

Another dwarf, a young woman if Owyn judged her appearance correctly, showed them to a room in the long hall. Gorath hesitated when he stepped inside. 'Something . . .'

'What?' asked Owyn.

'A feeling, of . . . call it a memory. Great power was once here.'

The young woman said, 'Lord Tomas used to rest here when he wintered in Caldara. I can sometimes feel it, too. If you need anything, just stick your head outside the door and call for me; my name is Bethlany.'

'Thank you,' said Owyn.

Owyn sat on a bed while Gorath looked at the other in the room. 'What they say of Tomas must be true, then, for me to sense the power of the Valheru ten years or more after he slept here.'

Owyn said, 'Anything is possible.' He lay down. 'But right now I need sleep.'

Gorath watched as the boy quickly fell asleep, but sleep was not something Gorath felt in need of. He left the room after a minute and walked to the door, then stepped outside.

Dolgan stood upon the porch of the long hall, looking out over the village. It was comprised of a dozen buildings of varying size, a few obviously dwellings, while the others appeared to be shops: a smith, a carpenter, a baker.

'Pretty, isn't it?' asked Dolgan.

Even without the flowers of spring yet apparent, the valley was a lovely place, nestled in pine and aspen. The people living there were industrious and everything in sight spoke of bounty. High enough up the hillside to be visible, cattle grazed in a meadow on the other side of a stand of trees. Chickens and ducks squawked as they hurried across the town's square, while a pair of dogs tried to herd them.

'It's a good place,' agreed Gorath.

'I've only seen a few moredhel villages, empty after the Tsurani drove your people from the high pastures. I remember them as not that different from here.'

'We build in a different fashion,' said Gorath. 'But shelter is shelter, and we bake and work the forge, much as you and the humans do.'

'I'm five hundred and twenty-eight years old next Midsummer's Day, and I've fought for my people for most of those years.' Dolgan looked up at the tall dark elf. 'Do you know that you're the first of your kind I've ever had a civil word with?'

Gorath sat on the steps. 'And I with a dwarf. Or a human until a few months ago.' He leaned back against a supporting post and said, 'I find the world a very different place than I thought it was when I was a boy. I was but twelve summers when the safety of my band fell to me, and I was thirty-seven summers when I avenged my father and became clan chieftain. For more than a hundred years the Ardanien tribes lived in the ice caves in the far north, where the sun never shines in winter and never sets in summer. We hunted seal and walrus, traded with tribes to the south of us, and lived apart even from most of our kinsmen.

'Then we returned and I fought to preserve my clan, and we rose and became a force within our nation. We had respect, we were feared, and when I spoke in council, the Ardanien were heeded.'

'What happened?'

'Murmandamus.'

'Which, the first or second?'

Gorath smiled. 'Both, you could say. The first was a remarkable creature. He spoke words that were compelling and insistent, and my people listened. I heard stories from those who had known him. We rose and struck south and overran the humans in Yabon.

'But Murmandamus died and yet his legend lived, and

when the second Murmandamus appeared, we were ready to follow without question.'

'Blind obedience is a dangerous thing.'

Gorath nodded. 'Before the second Murmandamus, some of my race were dislodged from the Northlands by more powerful clans, and they came south of the Teeth of the World. Others, like my clan, lived in the ice caves in the far north. We had one such upheaval a hundred years ago.'

'I remember,' said Dolgan. 'Some of your lads got a little bold and made free to come this way.'

'I have never before ventured so far south on this side of the Bitter Sea. When a lad I fished the sea near what the humans call Sarth.' He sat back and closed his eyes. 'I never thought I'd live to see the Grey Towers.' He looked at Dolgan. 'Some of my kinsmen, especially those who followed my cousin Obkhar, may be coming this way to live again in the Green Heart.'

'Well, as long as they stay down in the trees we won't trouble their passing. We never had much trouble with the Green Heart moredhel, but your clans up here in the mountains were not gracious neighbours.'

Gorath studied the dwarf and laughed. 'You sound like your son. As I told him, I suspect my people would have little charity in their description of you as neighbours.'

'Aye, that's true, I'm sure,' Dolgan chuckled. 'But what has long puzzled me is why that is so. We dwarves, despite our skills in warcraft, are a peaceful enough folk when left alone. We trouble no one who doesn't trouble us. We love our children, tend our herds, and winter in our long hall singing and drinking ale. It's a good life.

'But you're the first of your kind I've spoken with in peace since I was born, Gorath, so I must ask you this: why do you moredhel hate us dwarves and the humans so?'

Gorath considered the question for a long while, then said, 'When I fled south from my homeland, chased by my own cousin who sought to kill me, I would have answered you one way. I would have said, "When the Valheru left, they made us a free people, and gave to us this world, and you and the humans are invaders. You take what is ours".

'Now, I don't have an answer.'

'What's changed?' asked Dolgan, genuinely curious.

'Many things,' said Gorath. 'My own people have become . . .' He sighed, long and as if releasing something held back a long time. 'Many years ago we were much the same people, those of us who became the moredhel, eldar, eledhel, glamredhel. We were *the people* in our tongue. Most of our names were given by our enemies. Eledhel is a word that was coined by my people in contempt: the "elves of light" in the human tongue. It was a mocking name, hurled at those who sought to make themselves better than the rest of us. They called us "dark ones", or moredhel. We named the glamredhel, the "mad ones".

'We, who were once one race, are now so different, one from another, that I think we have lost any sense of what we once were.'

Dolgan nodded, but said nothing as he listened closely.

'Did you know that we cannot father a child on an eledhel or glamredhel woman?'

Dolgan shook his head.

'It is thought by our healers that something is needed between a man and woman of our race, something that has changed so profoundly we are as different now as dwarf or human to our own cousins.'

Dolgan said, 'That is most passing strange.'

'I am old by the measure of my people,' said Gorath. 'Two hundred and sixty summers will I see next Midsummer's Day. My birthright is three times that; only our cousins in Elvandar reach those spans of years, Dolgan. And that is because they have found one thing we have never known in the north: peace.'

Dolgan sighed. 'Peace is a wonderful thing to find, either

for one's people . . .' He looked Gorath in the eyes. 'Or within your own heart.'

Gorath looked out at the serene tableau before him and said, 'We live behind walls. Our villages are fortresses. No woman goes to herd sheep or cattle without a sword at her hip and a bow across her back. Our children play with weapons.' He hung his head, looking down at the dirt. 'We let them cut themselves so they learn early lessons. I despair for my people, Dolgan.'

Dolgan again was silent, then he said, 'I think you need to go to Elvandar. For more reasons than simply to take a message to Tomas.' He stood up. 'But right now I think you could use a long draught of ale. And I happen to know where we can find one.'

Gorath managed a slight smile and said, 'You treat an enemy with hospitality, Dolgan.'

Dolgan shook his head as he said, 'You're no enemy of mine, Gorath of the Ardanien. That's as plain as the beard on my chin.'

Dolgan led Gorath inside.

Owyn awoke to the sound of laughter and walked to the common room, to find Gorath and Dolgan and half a dozen other dwarves all drinking and telling stories. One of the dwarves not known to Owyn said, 'Aye, goblins will do that, if you convince them it's a good idea.'

Peering out the window, Owyn saw that it was morning and said, 'You've been drinking all night?'

Dolgan said, 'Welcome, my friend.' He put his feet down from where he had had them on the table and looked out the window. 'Aye, so it seems. Care to drain a flagon?'

'It's a little early for me, and besides, we must head for Elvandar.'

Dolgan said, 'True. Well, then, some food to break your fast, then on your way.' The old dwarf pounded on the table. 'Food!'

Soon the other dwarves had taken up the chant and were pounding the tables with their pewter flagons, shouting, 'Food! Food! Food!'

An old dwarven woman in a grey dress with her hair tucked up under a white linen cap entered from the kitchen, with a large wooden spoon. Waving it like a weapon, she said, 'Keep your armour on, you lazy louts!'

Half a dozen other dwarves followed, each carrying a platter of food. There were spiced fruits, hot sausages, loaves of steaming bread, jars of butter and honey and savoury flat cakes. And more ale.

Owyn sat down and said, 'I am astonished at how much ale you can consume without any ill effect.'

'A hearty constitution is a dwarf's heritage,' said Dolgan.

'Aye,' agreed Gorath. 'That's the truth. Try chasing one for three or four days.'

All the dwarves fell silent, then suddenly they all erupted into raucous laughter. Then with a wry, self-deprecating smile, Gorath added, 'Or running from one.'

The hilarity redoubled, the dwarves fell to the breakfast fare with vigour.

After the meal the horses were brought, and Owyn discovered they had been stocked with enough food for weeks. The animals had been fed and watered, and all the tack had been cleaned and repaired. Owyn said, 'Dolgan, my thanks.'

'For nothing, lad,' said the dwarven king. He pointed to Gorath. 'You gave me a rare chance to know this fellow, and it was my pleasure.'

Gorath extended his hand to Dolgan and they shook. 'Your hospitality is unmatched, friend dwarf.'

'And you are always welcome in Caldara, Gorath of the Ardanien.'

'I thank you,' said Gorath, and he mounted his horse.

A group of young dwarves approached, armed and armoured, and Dolgan said, 'I'm sending some of the lads with you to the River Crydee. They'll make sure you get there in good order.'

'Again, thanks,' said Owyn. They set out at a walk, with the dwarves moving out on foot. Owyn turned to Gorath and asked, 'You fit to ride?'

Gorath laughed, and said, 'No, but let's go anyway.'

'You are in an unusually cheerful mood, Gorath.'

'Yes,' said the dark elf. 'It's been too long since I've had the company of other warriors, good ale, and stories of valour and courage.' He lost his smile. 'Far too long.'

They were silent as they rode out of the dwarven village.

Travel through the woodlands of the Green Heart and the eastern edge of Crydee Forest was uneventful. A week after having left Caldara they reached the banks of a river. The leader of the dwarves, a warrior named Othcal said, 'We will part company here.' He pointed. 'That is the River Crydee. On the other bank is Elvandar.'

Gorath said, 'I could sense it since yesterday.' He spoke softly.

Othcal pointed down a trail. 'A bit more than a mile down there is the ford we use. Go there and wait.'

They bid the dwarves farewell, and rode on. 'Wait for what?' Owyn asked.

'You will see,' said Gorath.

They reached the ford, a large bar of sand held by stone which had caused the river to widen and run fast, but one which the horses could navigate without trouble. They waited. 'I don't mean to nag,' said Owyn, 'but what are we waiting for?'

'To be invited to enter. None may enter the elven forests unbidden.'

'What happens if you try?'

'Bad things.'

'I won't try. What do we do to let them know we're here?'

'Nothing. They know.'

A few minutes later a voice called from the other bank in a language Owyn didn't understand. Gorath replied in the King's Tongue: 'Two who seek entrance to Elvandar. We carry a message for Warleader Tomas from the Lady Katala, Pug's wife.'

There was a momentary pause, then a figure appeared on the other side of the river. 'I would know your name and line.'

Gorath said, 'I am Gorath of the Ardanien, chieftain of my clan.' He glanced at Owyn.

Owyn said, 'I am Owyn, son of the Baron of Timons.'

'Enter,' said the elf.

They rode their horses across the ford and halted as half a dozen elves appeared from behind the trees. The leader approached and said, 'We are a full day's ride to the edge of the elven glades, and another day to the Queen's court.' Without saying anything else, he set off at an easy run, while two other elves fell in behind. The remaining elves stayed behind.

Owyn studied them as he trotted along beside the elves, and realized he could not tell the difference between them and Gorath's people by casual appearance. But there was a subtle difference in their manner and bearing.

Gorath was tall, broad-shouldered and powerful. Owyn had seen him move, quick and deadly. These elves appeared more slight, less broad of shoulder and chest, yet equal to Gorath in height. But the biggest difference appeared to be how they moved. There was ease in their movement, as if they were one with the surrounding forest, and it was what Owyn could only label grace. They were graceful.

They ran for an hour, apparently without tiring, then halted to rest a few minutes. Gorath studied his distant kin and said nothing.

With some silent communications, the only part of which Owyn noticed was Gorath nodding slightly, the elves stood and waited while Gorath and Owyn remounted. They rode until sundown, then the elf who had bid them enter the elven woods said, 'We will camp.'

By the time Owyn had his horse unsaddled and tended to there was a fire burning in the clearing. A waterskin was passed and food appeared from hip packs. The elves sat upon the ground, or lay upon hip and elbow and remained silent.

After eating, Owyn spoke to the one whom he considered the leader, the first who had spoken and said, 'Might I know your name?'

'Caladain,' said the elf. He pointed to the other two and said, 'These are Hilar and Travin.' They inclined their heads toward Owyn in turn.

Owyn suddenly realized he didn't have any idea of what to say, so he remained silent. Gorath finally said, 'The eledhel aren't given to idle chatter like you humans.'

The elves smiled politely, as if they didn't feel quite the same way, but Owyn could see they were amused by the comment. 'I see,' was all Owyn said.

He finally got out his bedroll, spread it, and lay down without comment. Soon he was asleep under the bowers of the elven forest.

The journey continued with almost no conversation, but late in the second day, Owyn noticed the woodlands darkening off to his left. 'Is there something over there that's different from where we are now?'

Caladain asked, 'Have you some magic skills?'

'Yes, why?'

'Because most of your race would not notice the difference. Yes, that is one of the sleeping glades. Those who come here unbidden would be opposed by more than our warcraft. These very woods are our allies and we have many such places. In that stand of woods you would find yourself wanting to sleep and it is a sleep from which you would not awake without magic.'

Owyn glanced at Gorath and said, 'The bad things you mentioned?'

Gorath nodded. 'Our legends warn of many such dangers in the home of our – ' he glanced at his escorts ' – cousins,' he finished.

Owyn couldn't be certain, but he thought the elves looked troubled by the reference.

They moved across a tiny stream, and then up a rise, then entered a vast clearing. Owyn and Gorath reined in.

Separated from where they stood by an open meadow, a huge tree city rose upward. Massive trunks, dwarfing the most ancient oak, rose to stunning heights. They were linked by graceful branches, forming bridges that were flat across the tops. Most of the trees were deep green, but here and there could be seen one with leaves of gold, silver, or even white foliage, sparking with a faint light. A soft glow bathed the area and the sight of it warmed Owyn in a way he couldn't explain.

Elves could be seen moving along the branches, or at the base where fires burned as cooks laboured, smiths worked metal, and other crafts were undertaken. It was the most beautiful place Owyn had ever seen. He could hardly pull his eyes away, until Caladain said, 'Elvandar.'

Owyn looked at Gorath and saw his companion sitting in rapt amazement. His eyes were wide and shining, moisture gathering in them. He said something softly, as if to himself, in a language Owyn didn't understand. Owyn looked at Caladain, who said, 'He said, "How could we know?"'

'Gorath?' asked Owyn.

Gorath dismounted and said, 'It's a legend. Barmalindar, the golden home of our race.'

Caladain said, 'We will take your horses. Walk to that tree with the white leaves, and others will meet you and guide you to our queen.'

Owyn and Gorath moved across the clearing and as they neared the trees, they saw elven children playing. Elven women sat in a circle carding wool, and in another area elven bowyers and fletchers worked on bows and arrows.

Three elves approached and the first said, 'Welcome to Elvandar. I am Calin, son of Queen Aglaranna.'

Owyn said, 'Highness. I am Owyn Belefote, son of the Baron of Timons.'

'I am Gorath of the Ardanien.'

'What brings you to our home?'

'I bring a message from the Lady Katala, Pug's wife, to Tomas,' replied Owyn.

'Then follow me,' said the Prince. He sent one of the others ahead as he walked with Owyn and Gorath. 'You are the first of your people to come to us in many summers,' said Calin to Gorath.

A flurry of footfalls on the ground alerted them to a band of young male elves running after one who held a token. The one in the lead was blond, fair to the point of having almost white hair, and he was looking over his shoulder when he almost ran into Calin.

With a laugh, Calin caught him spinning him in a full circle, saying, 'Cautiously, little brother.'

The boy stopped and saw Owyn and Gorath, and said, 'Now I see why you speak the tongue of the Kingdom.' He stopped and said, 'Your pardon.'

'None needed,' said Calin with a laugh.

'We were playing hound and hare, and I was the hare.'

'You were on the verge of being caught.'

The boy shook his head. 'I let them stay close so they don't get discouraged.'

Calin said, 'This is Owyn, from the human city of Timons, and this is Gorath of the Ardanien.'

The Prince turned and said, 'This is my younger brother, Calis.'

The boy nodded and said, 'Welcome Owyn of Timons.' To Gorath he spoke in a different language, and at the end he seemed to be waiting. Then Gorath stepped forward and they shook hands. Calis looked over his shoulder at his friends, who were standing silently watching Gorath with intense curiosity. He shouted, 'Catch me!' and was off.

A moment later, the others were in pursuit. Owyn said to Gorath, 'What did he say to you?'

Gorath looked genuinely unsure of himself. 'He said, "I will fight you if I must, but I would rather you were my friend".' Looking at Calin he said, 'Your younger brother is a most remarkable youth.'

Calin nodded. 'More than you realize. Come, we have a short walk ahead.'

He led them up a flight of steps cut into the side of a huge tree. Calin warned, 'Don't look down if you have a fear of heights, Owyn.'

They moved deeper into Elvandar and the closer they got to the Queen's court, the more wonderful the place became. Soon they reached a large platform, upon which rested a half-circle of benches, and at the apex of the arc sat two thrones. Calin said, 'My mother, may I present two visitors: Owyn, son of the Baron of Timons, and Gorath, chieftain of the Ardanien.' He turned to the two travellers and escorted them to stand before a stunning woman who sat on her throne. 'My friends, my mother, Queen Aglaranna.'

The Queen was a regal beauty, with arching eyebrows atop wide-set eyes of pale blue. Her hair was reddish gold and she was serene in her ease. 'Welcome,' she said, with a musical note in her voice. To Owyn she said, 'Our human friends are always welcome in Elvandar.' To Gorath she said, 'As are our kin who come to us in peace.'

She motioned and said, 'Our ranks lack only your presence, Gorath.' He looked where she indicated and saw her advisors, a tall elf of many summers, next to whom stood one who was known to Gorath. 'Earanorn!'

The leader of the glamredhel nodded. His expression was cold, but he held his place. 'Gorath,' he said.

Another elf, one who looked as old as the first said,

'I am Aciala, of the Eldar, and am most pleased to see you here.'

Gorath was quiet for a long time, and Owyn was convinced some sort of communication was passing among the elves, silent but apparent to them. Then in a strange gesture, Gorath pulled his sword from its scabbard. He moved toward the Queen, and Owyn was suddenly alarmed. But he noticed no discomfort on the part of the others.

Gorath placed his sword at the Queen's feet, and knelt before her. Looking up, he said, 'Lady, I have returned.'

SIXTEEN

Tasks

The Queen stood.

She stepped down to stand before Gorath, then she leaned over and placed her hands on his shoulders. 'Rise,' she said gently.

Gorath did so and the Queen studied his face. 'When those of our lost cousins return to us, there is a recognition of this change within them.' Her smile was reassuring and her tone conciliatory as she said, 'But in you I sense something different. You have not returned to us yet, Gorath, but you are returning. Your journey back to your people is not yet complete.' She looked at the glamredhel leader and said, 'There are others here who also have not finished their journey, so you are not alone. When you have completed your return to us, then you will be given a new name, but until then you are still Gorath. But now you are Gorath of Elvandar. You have come home.'

She embraced him, holding him for a long, reassuring moment and returned to her throne. Owyn watched as Gorath picked up his sword and returned it to his scabbard. The young nobleman said, 'If it pleases Your Majesty, may I ask a question?'

'You may,' said the Queen as she sat upon her throne.

'I bear a message to your husband from the wife of Pug the magician.'

Aglaranna turned to Calin and said, 'Son, please escort these two to my private apartment.'

Prince Calin indicated Gorath and Owyn should follow him. They bowed once more to the Queen and she said, 'You may go, and when you've finished speaking with Tomas, return and we shall feast.'

As they walked away, Owyn said to Gorath, 'I don't quite understand what I just saw.'

'I'll explain it to you later,' said Gorath.

Calin said, 'My mother's husband was injured in a skirmish near the border with a band of moredhel who were trespassing on our lands as they raided south.'

'Those were not raiders,' said Gorath. 'Those were members of Obkhar's clan fleeing Delekhan for the Green Heart.'

Calin inclined his head at the correction. 'In any event, Tomas was struck by a poisoned arrow and is now resting.'

He pushed aside a richly-decorated tapestry and led them out onto a large private terrace that overlooked the splendour of Elvandar. In an alcove that could be masked off with a large curtain, a large man lay upon a bed. Calin said, 'Let me see if he's awake.'

'I am awake,' came a weak voice from the alcove.

Calin said, 'Tomas, this is Owyn, from Timons, and Gorath, one of our people returning. They bear a message from Pug's wife.'

Owyn and Gorath approached and saw a large young-looking man, easily six inches past six feet in height, lying beneath a down quilt, with pillows propping up his head. Gorath faltered as he reached a point a few feet from the bed. 'I had heard rumours,' he said softly, 'though few counted them true. But they are true. He is Valheru.'

Calin said, 'Not entirely, to our everlasting thanks.'

Tomas said, 'I would rise to greet you, but I am presently in little condition to do so.'

'Poison?' asked Gorath. 'What manner?'

'A thin green substance unknown to us.'

'Coltari,' said Gorath. 'It is rumoured to be from the Tsurani world, named after the province from which it came. It came to us about the time Delekhan began to rally the clans.'

Calin said, 'Is there any antidote?'

'May I see the wound?'

Tomas motioned for Gorath to approach and Tomas moved, turning his head to show an angry wound on the right side of his neck, just above the shoulder.

Gorath said, 'By rights you should be dead.'

Tomas smiled and Owyn was struck by his youthful expression. He was a striking man, with angular features, and his ears were almost as pointed as an elf's. 'I have discovered that I'm rather hard to kill. But I certainly can be brought to my knees. I haven't the strength of a day-old puppy.'

Gorath said, 'If he's lived this long, he will recover, but how long that will take I cannot say. Those who have suffered mild Coltari poisoning have lingered weeks before starting a slow recovery.'

Tomas said, 'I shall be myself in a few more days.'

Calin said, 'My mother's husband is always optimistic. He shall be abed for weeks, I think. Our healers have done all they can.'

'What is this message you bear from Katala?' asked Tomas.

Owyn said, 'She bids us tell you that Pug and Gamina have vanished from Krondor. Pug left behind a cryptic note: To Tomas! The Book of Macros!'

'We stopped at the Abbey of Sarth along the way and they have no knowledge of such a book. Is it something you possess?'

'Yes,' said Tomas, 'but it is not a book, in truth. Calin, would you bring me that box next to my weapons chest?'

Calin did as he was requested and carried a small box to Tomas. Tomas opened it and took out a scroll. 'Book of Macros is a phrase Pug and I agreed on to let me know he was in dire need of my help. He created this scroll so that whoever reads it will be taken to Pug.' He sat up. 'Calin, help me on with my armour.'

Calin said, 'No, Tomas. You have no strength. You will not help your friend in your present condition.'

'But Pug would only send such a message if he was in dire need of help.'

Calin said, 'I will go.'

'No,' said Gorath. 'We will go.'

Owyn said, 'Our mission from Prince Arutha was to find Pug, and if this takes us to him, then we have fulfilled our mission.'

Looking at Calin, Gorath said, 'No slight intended, Prince Calin, but if I am not a more experienced warrior than you I will be surprised. And you have a duty to the people here, to lead the warriors while Tomas heals.'

Owyn said, 'And I know some magic, Lord Tomas, which may prove important.'

Tomas said, 'Or I could wait a few more days.'

'Time is fleeting,' said Gorath. 'We have already been weeks on this journey, and soon Delekhan will launch his assault on the Kingdom. Arutha fears his magicians, which is why he wishes Pug's counsel. Let us go. We may not be the best choice, but we are here and willing.'

Owyn took the scroll and said, 'Please?'

Tomas nodded, and Gorath said, 'Tell the Queen we will feast when we return.'

Owyn unrolled the scroll and glanced at it. 'Gorath, stand behind me with your hands on my shoulders.'

The scroll was written in an alien tongue, yet the writing captured his senses and forced his eyes to follow line by line, and as he did so symbols appeared in his mind's eye, burning brightly like letters of fire. When at last he reached the final phrase, the area around them swam and flickered, then suddenly they were propelled through a glassy-looking plane hanging in a grey void.

Through a tunnel of light they raced, with sensations rushing through them like sounds and aromas, yet gone before they could be fully apprehended. Then suddenly they raced toward another shimmering pane of silver light and found themselves lurching forward onto the ground.

They stood upon grey powdery soil, with large reddish rocks rearing up on two sides. The sky was a vivid violet, rather than blue, and the air smelled of odd and alien odours. The wind was dry and cold.

'Where are we?' asked Gorath.

Owyn said, 'Nowhere on the world we know. We are somewhere else.'

'Where?'

'I don't know,' said Owyn. To the east a small and angry white sun was setting over the mountains, plunging the area into shadows of indigo and black. 'But it appears that night is falling and we need shelter of some sort.'

Owyn attempted to activate the spell he knew that permitted him to create light and suddenly he knew a terrible truth. 'Gorath! Magic doesn't work here!'

James studied the map. 'Are you sure?' he asked the soldier.

'Yes, squire. I saw at least three of their patrols moving along that goat trail and over the ridge.'

Locklear looked at the positions on the map and asked, 'What are they doing?'

'They can't move any significant number of warriors over that trail, so they must have been scouts. But scouting for what?'

'Perhaps they want to see if we're being reinforced?' suggested the soldier.

'Well, if they see reinforcements, I hope they have the courtesy to let us know,' said Locklear.

'It's if they don't see reinforcements that we should expect to be attacked,' said Jimmy, not seeing any humour in the situation. To the soldier, he said, 'Order a galloper to ride a day toward Dimwood and then back. I want word of any sign of the Prince being on the way. If not, I expect we'll be attacked soon.'

The soldier hurried out, and James turned to Locklear.

'I think that we have to assume Gorath and Owyn didn't make it.'

'So we must assume that we're holding this position until . . . ?'

'We get relief or we get overrun.'

Locklear was silent a moment, then asked, 'Do we organize a retreat like we did at Highcastle if it becomes apparent we're going to lose?'

James was silent a very long time as he considered the question. 'No. We stand or die.'

Locklear let out a long, theatrical sigh, and said, 'I guess that's why we have offices.'

'I believe Arutha would say, "It's time to earn our pay".'

Locklear seemed to muster resolve from within, and said, 'Very well, let's make sure we earn it with distinction.'

They left the late Baron's office and set about the many tasks before them in preparation for the coming battle.

The sun rose on a desolate and alien world. The few minutes the quickly-vanishing sun afforded them the evening before had been spent finding a shallow cave. It provided slight shelter, but they had spent a cold and hungry night in the dark.

As the sky lightened, Gorath woke Owyn. The young magician had been in a near-frantic state after discovering his magic wouldn't work on this world.

And that was the other shock that had almost reduced Owyn to mindless panic: they were on another world. There was no doubt about it. Gorath knew the northern night sky of Midkemia as well as anyone who lived much of his life out of doors, but even Owyn knew there were three moons on Midkemia, and not a single large one that bulked in the sky twice the size of the largest one at home.

'Where is Pug?' asked Gorath.

Owyn said, 'If the spell was designed to bring Tomas to him, he must be close.'

Gorath looked at the ground as the sky lightened above him. 'Look,' he said, pointing at the ground. 'Tracks.'

Owyn looked and saw scuff marks in the earth. 'Perhaps this was where Pug appeared, and that's as close as the spell can bring someone.' He winced as he considered what he just said. 'What am I babbling about? I have no idea what has occurred to us, or to Pug before us.'

Gorath knelt and studied the tracks. 'One pair of tracks begins there.' He pointed to a place roughly where they had appeared, then his hand traced a line. 'Whoever left those tracks went that way.'

He stood and began following the tracks while Owyn glanced around. The light appeared wrong, and the sky was almost clear, with a few wispy high clouds barely visible in the upper atmosphere. The wind was dry and cold and there was scant vegetation in sight, and most of it reminded Owyn of the high rocky hills through which he had travelled in the Northlands with Gorath.

Gorath said, 'Other tracks join here.' He pointed to a place on the ground that looked like nothing more than a few scuff marks in the soil. 'If those first tracks belonged to Pug, he was met or followed by at least four others. They all moved off in that direction.' He pointed to a series of small hills in the distance. 'Then I guess that's where we go.'

As the sun rose the day's temperature began to increase. 'This is a desert,' said Owyn. 'I've heard stories from those who've travelled in the Jal-Pur. The cold night fooled me.' He stopped and opened his travel pack. He removed an extra tunic, and tied it over his head, like a hood. 'Before we do anything else, we need to find water.'

Gorath looked around and finally said, 'You are right. I see no open water anywhere.' He looked at their apparent goal. 'I know little of deserts, but I would think if there was water to be found, it would as likely be found in those hills. Let us continue on that course.'

Without a better option to offer, Owyn agreed. They

trudged over a landscape of hardpan, broken rocks and eroded ridges. 'If anything ever flourished in this land, it has long since died out,' observed Gorath. As they walked, he asked, 'Have you any insight into why your magic will not work here?'

'No,' said Owyn, looking dispirited. 'I have tried every cantrip and incantation, meditation and exercise I could remember. All seem to function as they were designed to, but there was no . . . magic!' He shook his head. 'It's as if there is no mana here.'

'Mana?' asked Gorath.

'It's one term for it,' said Owyn. 'At least that's what Patrus called it. I don't know if other magicians do. But it's the energy that binds with everything else, yet can be manipulated to create magic. Most people don't understand how magic works. I don't have the power within me. All I know are series of words, actions, images; things that help me gather the power, or mana, from the world around me. It's as if the mana doesn't exist here. It makes me wish I was a Lesser Path.'

'What is that?' asked Gorath, letting the boy instruct him rather than go along in silence.

'They operate on a different path of magic. Patrus is a Lesser Path who suggested I go to Stardock when it was clear he was teaching me the "wrong" magic. Before Pug had travelled to the Tsurani world no one knew of the differences between the two paths of magic; at least no one on Midkemia.

'The Lesser Path is part of the magic, for lack of a better term. The magician links into the very nature of the soil below his feet, or the water that's everywhere, even in the sky, or the wind itself. The potential for most things to burn fuels the nature of those magicians whose nature is linked to fire. I think a Lesser Path magician might be able to milk some small magic out of this place, but I am powerless.'

'Would this be true of Pug?'

'I don't know,' said Owyn. 'He is rumoured to be more than Lesser Path or Greater Path.' Owyn glanced around. 'But it may also be that his powers are diminished enough that he might have been overcome here by others. I do know one thing.'

'What?'

'Without Pug we have no chance of returning to Midkemia.'

They walked on in silence for hours after that.

It was the middle of the next day when they saw the dome. The heat had reduced them to a slow, plodding walk, and there was no sign of water. The skins at their hips were now empty and Owyn was starting to feel the effects of dehydration. In silence they moved toward the dome. As they got close they saw it was a structure, made of some sort of hides stretched out over a web of sticks. 'It looks like a yurt,' said Gorath.

'What is that?'

'The nomads of the Thunderhell Steppes use them. They can erect them or take them down in minutes.'

He pulled his sword and moved around the building until he found the entrance, masked by a single curtain of leather. He pushed it aside with the point of his sword and when nothing happened, stuck his head inside.

'Come see this,' he said to Owyn.

Owyn followed him inside and looked around. The structure was empty, save for a faded cloth that once might have been a rug, upon which Gorath sat. He held out a piece of parchment. It was written upon in charcoal.

Owyn took it and read:

Tomas,
    As Katala no doubt sent you word, I imagine you have heard that Gamina and I are missing. She has been abducted by the Tsurani magicians who serve Makala and has been transported here. I will give you

more details once we meet, but I am limited to two pieces of parchment and must be brief.

Do not depend on Magic here. It does not function. I have some theories as to why, but again I will save those until we meet. Its absence may be due to this planet having once been visited by the Valheru, but your inherited senses may have alerted you to this fact already. A violent race resides here, and I have already had to elude four of them. They appear to be related to the Pantathians, similar enough in appearance that I judge they were left here by Alma-Lodaka during the time of the Valheru raids across the skies. Be wary of them, for I think they serve our enemies, somehow.

Gamina is still missing and I have made a sweep of the entire area. I am leaving in the morning to visit the northern tip of this island. From a hill nearby you can see what appear to be ruins there. Perhaps there I will find an answer. Look for me there.

Pug

'Well, that is where we must go next,' said Gorath.

'I wish he had said something about water,' said Owyn.

'If others live on this island, there will be water somewhere.'

Owyn nodded, but he didn't speak his concern, that they might not find it in time.

'At least we know it's an island, now,' said Gorath. 'That's good.'

'Why?'

'Because it means we can't wander for ever,' said the dark elf.

Owyn found the humour a little too grim for his taste and said nothing. They trudged on and when they crested the ridge, they could see in the distance the structures Pug had referred to. More, they saw what looked to be a vast blue body of water beyond.

'If we can get to the shore,' said Owyn, 'I think I can contrive a way for us to get water without salt.'

'Perhaps this island exists in a vast lake,' said Gorath hopefully.

'That would be welcome.'

They moved down the ridge and as they reached the bottom of the ridge, Gorath shouted, 'Water!' He bent and tasted it. 'It's fresh! Hand me your skin.'

Owyn did so and after a minute, Gorath handed him back the skin, partially filled, so that he could drink without waiting any longer. Owyn drank and Gorath said, 'Slowly. Too fast and you may swoon.'

Owyn forced the liquid from his lips. It was thick with the taste of minerals and warm, but it was the best water he could remember tasting. He watched as Gorath did the same with his own waterskin, then set it aside and took back Owyn's. This time he filled both skins and said, 'I will mark this place, so if we don't find another source on our search we can return here.'

Owyn nodded, and said, 'We are close to those ruins.'

Gorath said, 'We should reach there before sundown.'

They drank their fill, then moved on.

They encountered another yurt-like dwelling, almost buried in the dust, a few hundred yards shy of their goal. They had thought they had seen ruins, but at this distance, they could see seven massive pillars appearing to be fashioned from stone. Gorath again used his sword to move aside the leather curtain of the hut, and Owyn peeked in.

Inside he found another note:

Tomas,

While I have so far found no evidence of Gamina in the ruin, I have learned some things about this planet. Magic has been transformed here; what some magicians call 'mana' has been reduced to a crystal form. No natural phenomenon could account for such a transformation, so I can only assume some agency on the order of the gods did this, as even the Valheru would only have created a cataclysm by undertaking such a feat. It may be this act inspired Drakin-Korin to create the Lifestone, but that's a matter for us to ponder at our leisure.

I learned much by touching the pillars at the north end of the island. Avoid the centre one; I was ill for days after I touched it. In my weakened state I was almost overcome by two of the creatures I mentioned before. Only my skill with stone and sling saved me but the conflict taught me much. I have left an item for you; I do not know if it will help you with your Valheru-born magic, but I thought it would do no harm to leave it for you. Perhaps after I have found Gamina, I will have an opportunity to study more of the wonders on this world.

Pug

Owyn looked around and saw a long bundle set to the side of the round tent. He unwound another faded rug, identical to the one found in the previous hut, and inside saw what looked to be a staff fashioned of a strange blue crystal. He touched the staff, instantly snatching his hand back.

'What is it?' asked Gorath.

'I am not sure,' said Owyn. He slowly reached out and touched it again. 'This is amazing,' he said.

He held out his right hand while he touched the staff with his left, closed his eyes, and a moment later a glow of light emanated from his fingertips. 'I can't explain it, but this staff has given me back my powers. It's as if this staff is made of . . . I don't know . . . the crystallized mana Pug spoke of.'

'Bring it,' said Gorath. 'We should make for the ruins before we lose the light.'

*   *   *

They stood on the point of the island, a high bluff which overlooked an alien sea. Seven giant pillars of crystal rose up to seven times a man's height. Owyn said, 'I'll start with that one.'

He went to the pillar farthest to the left and touched it.

Despite a rocky appearance, it felt smooth to the touch. He squinted his eyes and saw he was actually running his fingers over a sheath of energy that clung to the surface of the pillar.

Owyn looked into the pillar and its many facets refracted images of the desert, sea and sky, but he also glimpsed other places, as if the pillar showed him different lands, oceans and skies.

Your observations intrigue me. You are savani, *are you not?*

Startled at the presence of an alien thought in his mind, Owyn shook his head. Unsure if he could simply think his reply, or speak it aloud, he decided speech would keep him focused. 'I am not familiar with the term *savani*, so I don't know if I'm one or not. With whom am I communicating?'

Gorath looked at Owyn with surprise on his face. Before Owyn could say anything to him, the voice returned to his mind. *I am Sutakami, Mother of the Thousand Mysteries, once a goddess of Timirianya. You have awakened me. What do you desire?*

'I'm not sure what you are asking me,' said Owyn. 'Are you an oracle?'

No. I may only tell you that which is already known, although I dimly sense things that may come to be. I sense you are new to this world. Perhaps you would wish to know of the creatures who inhabit it.

An image filled Owyn's mind before he could speak. The race was proud in appearance, like magnificent birds with arms instead of wings. Beaks were small and looked as if they could articulate speech. *These are the Timirian. They were poets and scholars, and warriors of great skill. They were on the verge of spanning the stars when the Valheru came. They were obliterated.*

Then another figure came into his mind, a shadowy creature of impressive aspect, the very features causing Owyn to flinch. Although a huge set of wings dominated the figure, it was the eyes of the creature, cold icy orbs of blue, that held Owyn's attention. *These are the ancient servants of Rlynn Skrr, the last High Priest of Dhatsavan, our Father of Gods, before the Great Destruction. Creatures of magic, they now wander free of fetter, so flee if you see one, for they may be killed only by a magic designed to drain their energy into the soil. Now they wander the ancient ruins of the temple of Dhatsavan.* The voice faded and grew distant. *I must rest . . . I am needed elsewhere.*

'Wait!' Owyn put his head down, as if tired. 'I need to ask more.'

Gorath asked, 'What is it?'

'These pillars, they're . . . ancient gods of this world. I was speaking with this one, a goddess named Sutakami.'

'Perhaps if you touch another?' asked Gorath.

Owyn nodded. He moved to the second pillar and touched it. 'I wonder what this place was originally,' Owyn asked.

You stand in the ruins of the Temple of Karzeen-Maak, once the high temple of the seven gods of Timirianya. Once, these columns were but symbols of the gods, crafted by the savani artisans who were the servants of Dhatsavan. Now they are the vessels within which we have taken refuge.

'What could drive a god into refuge?' Owyn wondered.

The Valheru, came the instant reply. *They extinguished life as we knew it on this world, leaving behind little. Only when Dhatsavan showed us that our struggles were futile did we create a plan to rob the Valheru of power, driving them from our world. They fled lest they be trapped here, leaving only a few of their servants behind.*

'What did you do?' asked Owyn.

Of the Seven Who Ruled, only six of us survived the Great Destruction. Two have faded so far from the world they can no longer give voice to their thoughts; they are now but sentient forces of nature. Only Dhatsavan will remain, waiting for the time of the

Awakening. He shall call us when the need has come . . . We shall not speak again, savani.

Owyn looked at Gorath. 'The Valheru caused this desolation.'

Gorath said, 'They were a power matched by few. Our legends tell of them spanning the stars on the backs of dragons. Only the gods were greater.'

Owyn looked around as the sun began to set. 'Apparently not all the gods. These pillars are what is left of the seven most important gods of this world. One is dead. Two of them are mute, two I've already spoken to.'

'Pug's note said to not touch the centremost.'

'So that leaves one more with whom to speak. Perhaps I can find out what happened to Pug from it.'

Owyn touched the next pillar, but was greeted only by a faint sensation; nothing of coherent thought. 'This must be one of those who has faded to mindlessness.'

He walked past the centremost of the seven, heeding Pug's warning, and went to the next pillar.

He touched it and found it lifeless. Not even the faint sensation he had noticed on the last one remained. He moved on to the next one.

Touching the pillar, still warm from the afternoon sun, he wondered who the Valheru had left behind.

The Panath-Tiandn. They are creatures from another world, trained to act as artisans of magic. They have limited intelligence, but they are clever, and dangerous. They created artifacts for the Valheru.

'Did they capture Pug?'

No, though they thought to, but I prevented it.

'Who are you?'

We seven were the gods of this world, and I, savani, was once Dhatsavan, Lord of the Gates. But when the Valheru brought their wars of desolation, we chose these forms rather than risk the final death.

'I don't know the significance of much of what you say,'

said Owyn. 'I have heard legends of the Valheru on my home world – '

What you know is unimportant, said the voice in Owyn's mind. *What we were is lost in time, but there is time for you to save your people from our fate.*

'Our world?' said Owyn. 'The Valheru have been dead on my world for ages. They can't pose any threat to us.'

A feeling of disinterest washed over Owyn, as if what he was saying was of no consequence to this being. *The one you know as Pug of Stardock will tell you more when the time comes for you and your companion to make your choices. For now, you must bring to this place the Cup of Rlynn Skrr. Do this and we will free Pug from his captivity.*

'What do gods need of mortals to fetch and carry for them?' demanded Owyn.

A sense of amusement came over Owyn as the voice replied, *You are wise to question, young savani, but it is for me alone to know the truth. Seek the cup in the far caves on the southeast corner of the island. You will have to kill the Panath-Tiandn who has it. Bring it to me or else perish in the desolation of Timirianya. The choice is yours. I warn you, do not attempt to use the cup. Pug has already learned the harsh lesson of trying to utilize its power without my guidance. Go.*

Owyn said, 'We must fetch a magic item from the far side of this island. And it seems we must battle some creatures of the Valheru to do so.'

Gorath said, 'It's been a long day. Let us return to that tent down the way and rest. A little food and sleep will help prepare us.'

Following Gorath, Owyn hoped that was true.

It had taken half a day to find the corner of the island where the frozen god had told Owyn they'd find the cup. Now they rested on a ridge above what looked to be a village, or at least a collection of huts in front of a large cave.

They had been watching for half an hour or more and seen

no sign of movement. 'Well,' said Owyn. 'Maybe they're deserted.'

'No,' said Gorath. He pointed to a pile of firewood. Then he pointed out a set of covered urns. 'Water, I think.' Then he pointed to what could only be scraps of food thrown into a trench near the edge of the village. 'There may not be many of these creatures left on this world, but this area is not abandoned.'

'Well, maybe they've all gone off somewhere.'

'Or maybe they sleep during the heat of the day and they're all inside?' suggested Gorath. He stood up. 'We won't know until we go down and see what is there.'

Owyn followed the dark elf down the hillside and when Gorath reached the first tent, Owyn said, 'The cup is in that cave.'

Gorath had taken one step when the leather covering of the hut he was about to enter swung open and a creature started to emerge.

Owyn's very skin crawled at the sight of it. An upright lizard, swathed in dark clothing, stood blinking in the sunlight. He had no opportunity to raise alarm, as Gorath thrust with his sword, running him through.

'Three,' said Gorath.

'Three what?' said Owyn.

'There are three more left if this is one of the four who were tracking Pug.'

'Or there may be a dozen left, if they're not the ones,' whispered Owyn. 'Let's be quick!'

They hurried to the cave and as Owyn started to push aside a large curtain hung across the entrance, it moved. He jumped back as a serpent-man hurled himself at Gorath. Gorath barely avoided a club strike to the head, and dodged back.

Owyn turned as another serpent creature snarled and leaped upon him, knocking him back. Owyn rolled on the ground, barely hanging onto the staff of crystal. The creature's face was painted with yellow symbols, and Owyn knew he was struggling with some sort of Panath-Tiandn shaman. Owyn saw claws above his face and locked gaze with the creature.

Suddenly symbols of fire burned in Owyn's mind's eye and he sent out a mental blast which rocked the creature back. Owyn spun out from under it and jumped to his feet. The creature struggled to recover from Owyn's assault. Owyn kicked the creature as hard as he could in the head, and it collapsed.

Two other serpent-men appeared as Gorath killed the one he faced. Owyn reached into his memory for another spell and felt the staff grow warm in his hand. A sphere of fiery energy exploded from his hand and struck the first creature, engulfing it in flames. The second was splashed by flame and its robe was set afire.

The first fell to the ground, dying in seconds, but the second fell and rolled, screaming as it tried to put out the flames. Gorath hurried over and put it out of its misery.

Owyn looked about and waited to see if any other of these creatures were present. All was quiet.

Finally, Gorath put up his sword and said, 'Let's find this damned cup.'

Owyn went inside the dark cave, illuminated by only a single flame in a tiny brazier, and his skin crawled. The place was a centre of dark magic and while he couldn't read the symbols painted on the walls, the shapes were alien and he could sense their evil. He glanced around and saw what looked to be a small shrine. Upon it rested a cup carved out of some unknown stone.

He reached out and took it, feeling energy rush up his arm as he gripped it. Once outside the hut he said, 'This is it, no doubt.'

'What does it do?'

'I don't know, but I was told that it harmed Pug, and if that's so, I will not risk trying to unravel its mystery.'

'Then let's get it back to those so-called gods and see if they live up to their part of the bargain.' Gorath looked around. 'I doubt these are the only members of this tribe on this island, and when they see what we've done, I think they'll be on our trail.'

'Can we reach the pillars by sundown?'

'If we start now and don't stop,' said Gorath, turning and setting off without waiting to see if Owyn was with him.

Owyn hesitated a moment, then set out after Gorath.

SEVENTEEN

Misdirection

Cold winds swept the battlements.

James signalled the archers to be ready to offer covering fire to the approaching horsemen. Two scouts raced up the incline to the drawbridge, whipping their lathered horses to get them to the gate before it rose too far to reach. James hoped he had timed it right because too early and those riders were stranded outside the walls; and too late and enemy riders might gain the barbican, and with the small band of defenders at his disposal, any enemies inside the castle posed a serious threat.

The first rider reached the bridge as it started to rise, and the second had to kick his horse hard to make it leap aboard the rising bridge, but they made it as James gave the order for covering fire.

Bowmen launched a flight of arrows at the pursuers, who fell back as three in their van were knocked from their horses. They were almost entirely human renegades, with two moredhel horsemen in the mix. They milled around just out of bow range, until James gave the signal for the firing of a single catapult, which showered them with stones, killing another half dozen.

The rest retreated down the road from the castle.

James was down to the barbican before the cheering on the walls had faded, asking, 'What did you see?'

The lead rider, a young corporal, said, 'No sign of help to the south and too damn many of the enemy coming down from the north.'

'What does it look like to the north?'

The young corporal rewarded James's faith in one so young by reporting calmly, despite the close call. 'Screening cavalry, who didn't take kindly to us pokin' about, squire. I could see a lot of dust and some of those siege engines you told us about in the distance. Looks like they'll be at the base of the road before nightfall.'

'You did well,' said James. 'Go get something to eat and some rest. We're going to have a busy morning.'

James went looking for Locklear, whom he had placed in charge of stores and weapons. When James found him, Locklear was in the middle of one of the storerooms looking disgusted at what he had found inside a barrel.

'What is it, Locky?'

'The meat's full of maggots. I think those Nighthawks got down here and did some mischief before they set about killing the officers. They didn't want the men to have a lot of reason for staying here, I think.'

'How bad is it?'

'All of the stored meat is bad. Most of the flour has bugs. We can sift those out, I guess, but I wouldn't want to be eating the bread unless I was starving. The hard bread looks all right, and most of the dried fruit is still edible. We can last a while.'

'I don't think food is our worry.'

Locklear looked at James. 'They're coming?'

'Tomorrow.'

'Then we'd better be ready.'

James nodded. He knew that he could expect the best of the men; they were all veterans of the border wars, but none of them had been tested in a full-blown defence of the castle. He knew the theory: he had studied with Prince Arutha, and he knew the reality, he had fought at Armengar and Highcastle, and he knew that the attackers needed ten men for each of his defenders on the wall. What had James fretting was his concern about what would happen if Delekhan brought more than ten to one against his position.

*   *   *

Owyn carried the cup to stand before the column. He touched it to the crystal spire.

Dhatsavan's voice sounded in his head: *You have returned with the cup. That is good.*

'Why do you need this?' asked Owyn.

I do not need it. I needed to keep it out of the hands of the Panath-Tiandn.

'Why?'

It is many things, an item of immense power, but one of its uses is that of a key. It allows access to other worlds. The abandoned children of Alma-Lodaka will be confined to this world for the time being. They are by themselves nothing more than a nuisance. Under the guidance of their Pantathian cousins, they are a dangerous tool. Eventually someone may fetch them from our blasted world, but for now the rest of the universe is safe from them. Take the cup with you and keep it safe.

'Pug's welfare is our concern. We have fetched this cup for you from the other side of the island. Where is Pug?'

He is safe within a structure constructed by the Panath-Tiandn. The protective barriers that keep him isolated within that structure will be removed once you locate him. He misapprehended the scope of the cup's powers. When he awakened its powers to seek the mind of his lost daughter, it overwhelmed him and reduced him to little more than a helpless child.

'You imprisoned him to protect him?'

The former god's reply was tinged with amusement, though Owyn wondered if human terms did justice to what he felt. *As an individual he is of little interest to us, but he was useful in preventing the Panath-Tiandn from possessing the cup for a while. They had been involved in a long process of unravelling the mystery of the cup and were close to understanding it. Pug interrupted that process and set them back years. That alone warranted our thanks. Now that you are here, we can see the cup gone from this world, and as payment for your service, we shall grant your friend's freedom.*

Pug has regained most of his identity and memory, but his abilities will yet be impaired for many days to come. Go to a hut to the west of the one in which you found the cup; there you will find Pug.

'How will we return home?' asked Owyn.

The way is now opened to a place in the mountains, caves that lead to a cavern in which the Valheru dwelled. Take Pug from the hut and to the north you will find an entrance in the mountains; there you will find artifacts that will aid you in returning home. Use the cup to teach Pug what you know and take the cup with you for safekeeping. Suddenly the knowledge of how to use the cup came to Owyn. *Then seek his daughter in a place near the mountains, where the Panath-Tiandn guard her, thinking she is an omen from Alma-Lodaka. Free her and return to your own world. But do not tarry, for I can only keep the gate to your world open for a limited time. My powers are not what they once were.*

Go now.

'Thank you,' said Owyn, and he motioned for Gorath to accompany him.

'Where now?' asked the dark elf.

Pointing back the way they had come, Owyn said, 'From where we found this cup we head west, and there we will find Pug. And when he is free, we find our way home.'

Gorath said, 'Then let us hurry. I tire of this harsh and desolate land.'

Owyn agreed.

James raced up the steps to the wall as the bugles sounded. Drums thundered outside the walls and he heard the sound of crossbows and short bows being fired before he crested the battlements. Locklear shouted, 'They're coming up from the north face!'

James nodded and glanced eastward and saw the large siege towers being rolled up the road. He hurried to the north wall and saw goblins climbing up the slope of the hillside below the wall, all of them carrying coils of rope and grappling hooks. Slightly smaller than humans, the goblins were almost comic figures when they weren't trying to kill you, James thought.

Black hair formed a heavy thatch above thick brow ridges. Their skin was blue-tinged, as if a fair-skinned human had been lightly stained with dye, and their eyes were black irises on yellow. They carried small buckler shields on their arms and short swords on their hips.

Defenders began shooting at the goblins, who started moving in shifts. Crawl a few feet upward, raise the small shields over their heads, then as soon as a shield was struck they scampered a few more feet upward.

James shouted, 'Bring rocks!'

Immediately soldiers came pulling shallow waggons on wooden wheels that carried large stones, ranging from the size of a man's fist to the size of huge melons. Soldiers slid poles under the carts and levered the carts upward, with one holding a large rope handle on the side of the cart.

The contents of the cart spilled over the wall, showering the goblins with stones, effectively scraping them off the cliff face. Screams from below showed the efficacy of the defenders' response.

'This is a diversion,' James said. 'Locky, check the other two walls while I go to the gatehouse.'

Locklear hurried off and James ran along the pallisades toward the gatehouse. He knew it was going to be a long battle, one without quarter. If we can just make them retreat, he said silently to himself as he hurried to the gates.

Gorath approached the hut with caution. They had been attacked by three of the Panath-Tiandn en route, and Owyn had depleted the crystal staff. He had been forced to hold one at bay by clubbing it while Gorath killed the other two, then dispatched the last.

Gorath yanked back the sword with a grimace of pain. 'What?'

'There's a barrier at the door. As soon as I touched it with my sword I felt a shock shoot up my arm.'

Owyn hesitated for a moment, then removed the Cup of

Rlynn Skrr from his belt pouch and tentatively touched it to the door. He felt power surge into the cup and saw it flicker a moment, as if illuminated, then nothing. He pushed aside the curtain and entered.

Pug stood in the middle of the room, looking disoriented. He blinked at the light and asked, 'Tomas?' He tried to rise, with the aid of a crystal staff, the twin of Owyn's, but sat back down after a feeble effort.

'No,' said Owyn as he entered, with Gorath behind. 'Tomas was injured in an attack on Elvandar. He recovers from a poisoned wound. We came in his stead.'

'Who are you?' asked Pug. Then his eyes narrowed. 'Wait, I remember you. You're the boy who came to Krondor with Squire Locklear, months ago.'

'Yes, and do you remember Gorath?'

Pug nodded. 'The moredhel my daughter tried to read.' Suddenly his eyes widened. 'Gamina! I must find my daughter.'

'We know where she is,' said Owyn. Lowering his voice he said, 'More or less.'

Pug seemed disoriented. 'I am weak, but my memory has been returning.' He looked at his right hand, which Owyn noticed bore a nasty old scar across the palm. 'But my powers have fled and I remember almost nothing.' He looked at Owyn and Gorath. 'You claim you come in Tomas's stead, but how can I be sure you're not here on behalf of our enemies?'

Gorath looked incensed. 'You accuse us of being false? You think us spies?'

Pug said, 'I only know you were the first tool of Makala.'

'Makala?' Owyn's expression was confused. 'The Tsurani magician?'

'This was his plan,' said Pug. 'I'm not saying Gorath is a willing tool, but he was as much a part of Makala's plan as Delekhan is.'

'Delekhan is also a tool of this Makala?' asked Gorath.

'I believe so. When you first brought word of Delekhan raising the war banner of Murmandamus over Sar-Sargoth, I became alarmed. Having seen Murmandamus's dead body with my own eyes, I knew he no longer lived. But I thought it possible the Pantathians were responsible, using a rumour of Murmandamus's survival as a goad to once again rally the moredhel to try to seize Sethanon.

'I used my abilities to get what information I could, as did agents of the Prince, and between our efforts we realized there was no link between the Pantathians and Delekhan. I then judged Delekhan but a simple warlord seeking to seize power for himself under the guise of seeking to free Murmandamus.'

Pug looked weak, and Owyn said, 'We have water and some food.' He offered water to Pug, who drank deeply.

Pug waved away the food. 'Later. Something Gorath said when speaking to my daughter nagged at me, and now I realize there was a link before me that had been apparent had I but the wits to notice.'

'What was that?' asked Gorath.

'You said that Delekhan had displayed the helm of Murmandamus as proof he still lived.'

'Yes, the dragon helm, black with lowered wings on both sides of the head.'

'But last I saw that helm it lay in a basement below the keep of Sethanon in an ancient stone hall,' said Pug. 'By no arts I know could Delekhan have reached the place where that helm lay. Someone else had to fetch it and return it to him.

'There are only four I can think of who might have the powers to discern the location of that chamber, and be able to find a way within: Macros the Black, myself, Elgohar of the Assembly, and Makala. Macros has been missing since the end of the Riftwar; Elgohar has served me well and has been busy with students at Stardock, leaving only one other.'

'Makala,' said Owyn. 'But why is he doing this? I mean it

explains one part of this puzzle that had frustrated Squire James, the part played by Tsurani in all this – '

'That was what alerted me,' said Pug. He finally stood, shaking a little. 'When nothing tangible is apparently being gained in a transaction, one must assume something intangible is being exchanged.'

'Information,' said Gorath.

'And service,' said Pug. 'I am now certain the Six you have spoken of are Tsurani Great Ones under Makala's direction. He admitted as much.'

Owyn asked, 'But why is Makala pushing Delekhan into a war with the Kingdom? Is it revenge over the Riftwar?'

Pug was silent a minute while he framed his answer. 'What I tell you touches upon some of the most vital interests of Midkemia, not just the Kingdom.

'When the Battle of Sethanon raged, Tsurani soldiers came to help, as did two friends of mine from the Assembly on Kelewan, Hochopepa and Shimone, and it's obvious that despite the usual Tsurani reticence, gossip of the final events of that battle reached certain ears on Kelewan.'

Pug took a deep breath, as if telling the story took strength he didn't have. 'Deep under the city of Sethanon lies an ancient chamber.' He put one hand around his other balled fist to demonstrate. 'But it is really two chambers in one space, one out of time with the other.'

Owyn's eyes widened. 'Shifted in time? Only the most powerful of magicians could even conceive of attempting to reach it.'

Pug nodded. 'The first chamber, in our time, is where Murmandamus died, and there Makala would have found the helm he gave Delekhan.

'The other chamber, however, the one that is always seconds ahead in time, is his real goal. It contains an artifact of immense power, a thing so dangerous it could spell the end of all life on the world of Midkemia. A mortal could stand in that chamber until the end of eternity and never "catch up" with the objects placed in the time shift; he would always be a few seconds too late to see the artifact. And that is what this war is about; it's a diversion on a massive scale to allow Makala to establish the spell he needs to shift time to get into that chamber.'

Gorath said, 'What is the need? Why send thousands to needless death to reach this second chamber if it is in the same place as the first? If he is a magician of such mighty arts, why not use his skills and slip into this other time using stealth?'

Pug said, 'I have studied this object for nearly ten years and have only begun to gain an inkling of knowledge as to its nature and purpose. In the wrong hands, it could wreak havoc undreamed of in our worst nightmares.

'Because it is so vital that no one reach this artifact, I erected additional defences around it. As I said, it is shifted in time, an act of the Valheru that I have left in place.

'And,' said Pug, 'secreted within the vast cavern is an ancient dragon, a guardian oracle of special abilities. Even my powers would be sorely taxed to best this remarkable creature, and if any agency should threaten this dragon, she would call to the King, who has placed a special garrison near Sethanon, stationed in the Dimwood against just such a risk. I am certain someone of Makala's intelligence has discovered that force and I believe he intends to use Delekhan's forces to attack that garrison so he can gain entrance to the lower chamber without soldiers coming to the dragon's aid. Even if the dragon is vanquished, Makala and his confederates will have their hands full in reaching the time-shifted chamber. They could not hope to do so while Kingdom soldiers were attacking them. They need many hours' preparation once they best the dragon.'

'The dragon!' said Owyn. 'It's the one I spoke with at Malac's Cross!'

'Yes,' said Pug. 'The old statue is used to contact the oracle, so that any who might come looking for her will be led astray. If you spoke with her, your mind was at Sethanon.'

Owyn looked at Gorath. 'Then that would be why she said you would play a critical role in this.' Looking at Pug, he said, 'And it also explains Delekhan's plan! Prince Arutha sent us to find you because he fears Delekhan will employ magic in his attack on Northwarden. James thinks once they've come through Northwarden, Delekhan's army will use boats to go downriver to Romney, then overland to Sethanon. Can you stop them?'

'In my present state, no,' said Pug. 'Even as we speak I am regaining memories and some of my physical strength is returning, but I fear it will be some time before all my powers return. In my blind haste to find my daughter, I used a magical artifact that promised to impart knowledge. But I would have been better served to have avoided it.'

Owyn nodded. He reached into his pouch and again extracted the Cup of Rlynn Skrr. He held it out before Pug. 'The being who called himself Dhatsavan said that this which robbed you of your ability could return it, but that I must aid you.'

Pug reached out and tentatively touched the cup held by Owyn. Owyn felt a tingling in his fingers. Images, feelings, unfamiliar memories, a sense of power, all rushed into his mind. Softly Pug said, 'This is a risk, Owyn, and in days to come you may find you have undertaken a burden you didn't anticipate. But for the moment, it will aid me greatly.'

Then Pug and Owyn slipped into darkness.

Owyn and Pug both roused at the same time as if from a deep trance, and found Gorath sitting next to them. 'I had begun to fear you would never awake,' he said as he helped them both to sit up.

'How long were we out?' asked Owyn.

'Two days,' said Gorath. 'You were in a trance and if I

put food or water to your lips you ate and drank, but you otherwise sat immobile holding the cup.'

Owyn blinked and felt images and ideas swimming around in his head. If he focused on an object in the hut or concentrated on a subject, they faded; but if he tried to relax, the fragments of thought again swirled through his brain. He stood and felt dizzy.

Pug stood and took the crystal staff. He stared at his hand and a flame erupted from the palm. 'Interesting. I could never do that before.'

Owyn said, 'It's a trick I learned from a magician named Patrus.'

Pug said, 'I seem to have abilities new to me, while those that should be familiar are just outside my grasp.'

'And I have new and strange images in my head that I cannot quite grasp, either.'

'In time many things may manifest themselves to you and should you need aid understanding them, come to me,' said Pug.

Owyn looked at Pug's staff and said, 'Mine has lost its magic ability, it seems.'

Pug said, 'We have to find you some more of the crystal that is the essence of magic, mana as some call it.'

'I thought the staff was mana.'

'No, come and I'll show you.' He led them outside and looked around the alien landscape. Most of the plants were fibrous and tough, with growths on them that looked like frozen crystals. 'That one over there,' he said. A single plant stood in the midst of others, but it was a golden colour while the others were purple or blue. 'This is not really a plant,' said Pug. 'Touch your staff to it.'

Owyn did so and saw a tiny flash and the plant vanished. He felt a thrum of power in his staff. 'Look for the golden plants as we walk,' said Pug. 'But for now, let us find my daughter and return home.'

'If it is not already too late,' said Gorath.

'Dhatsavan must have known how long this would take,' said Owyn. 'If we needed to flee before making this transfer of abilities, he would have warned us.'

Gorath nodded. 'We can hope that is true.'

Owyn pointed. 'He said straight north of here we would find an area utilized by the Valheru when they warred on this world. Near there we shall find your daughter. He said the Panath-Tiandn view her as holy and will not harm her.'

'That's a blessing, if true,' said Pug, the relief on his face obvious. 'Let us go.'

They hurried northward as the day wore on and stopped only once to drink water and rest. Owyn saw several golden plants and touched his staff to each, charging it with magic.

Near sundown they started to hear an odd, low sound coming from the north. As they got closer, the sound grew louder. They reached a ridge and found a half-dozen of the serpent creatures in a circle, with another dozen arrayed beyond, all bowing to a large hut with mystic symbols painted outside.

Gorath said, 'This will be difficult, especially those two on either end with staves like yours.'

Owyn said, 'A moredhel spell-caster named Nago tried to freeze me with a spell; I've made it work once.'

Pug closed his eyes and said, 'I . . . know which one you mean. The magic fetters that inflict damage. I . . . think I can cast that.'

Owyn said, 'If we can immobilize those two, then cast a ball of fire at the rest, maybe that will cause enough panic we can get inside and find your daughter.'

They agreed on a plan of attack, and when Gorath gave the signal, Pug and Owyn stood and, gripping their crystal staves tightly, incanted spells which sped across the clearing and struck the two alien spell-casters. Both were gripped by forces which froze them in place and inflicted terrible pain on them as energy crackled in the falling evening light.

The other Panath-Tiandn were stunned by the shock and stayed rooted long enough for Pug and Owyn to cast balls of fire into their midst. Several of the lizard-men shrieked and ran, their burning robes spreading flames. Others turned to the source of the attack and Gorath was among them.

Owyn used his blinding spell to stop one, while Pug cast another bolt of the evil purple energy which froze its target.

Gorath cut down the first Panath-Tiandn who faced him, and turned as another swung at him with a sword. He blocked the blow and turned it away, dancing backward and getting ready for another attack.

Two of the lizard-men turned to flee and Pug and Owyn struck at the others with their crystal staves. The cudgels proved surprisingly sturdy, delivering sharp blows without shattering. Soon the lizard-men were either dead or in flight.

Pug raced forward to the hut and threw aside the oddly-woven tapestry door. In the middle of the hut rose a statue, roughly woman-shaped, but ancient and without detail. 'Where's Gamina?' Pug asked aloud.

'Perhaps in the ancient Valheru stronghold?' suggested Owyn as he looked into the tent.

Gorath entered and said, 'This place is already heavy with their essence.'

Pug looked around for anything which might point the way to his daughter.

Gorath picked up a dusty bundle in the corner, obviously not moved for a very long time. Underneath he found a suit of armour, white with crimson-and-gold trim. Gorath dropped it as if it were burning. 'Valheru!' he exclaimed.

Pug touched it and said, 'Yes, it is very much like the armour Tomas wears.'

Owyn said, 'Is it dangerous?'

Pug ran his hand over it and after a moment said, 'No, there is no spirit of the Valheru within it. I think that quality was unique to Tomas's armour.' To Gorath he said, 'It is, however, astonishingly durable and nearly impregnable. Why don't you take it?'

Gorath shook his head emphatically. 'No. I have no desire to wear relics of my people's former masters. The desire for those trappings is a large part of why so many of my kin have walked the Dark Path. It is that lust more than anything which has kept my people mired in savagery while our eledhel kin have achieved a grace my people can't even begin to imagine.'

It was the most impassioned statement Owyn had heard from the dark elf since they had met.

Pug found an ancient-looking scroll and unrolled it. 'Look at this, Owyn.'

The young magician came to stand behind Pug and began to read over his shoulder. 'What is it?'

'I'm not sure, but I think it's the spell Dhatsavan told you of, the one which drains strength from those creatures of Rlynn Skrr.'

Owyn continued to read, and said, 'I feel . . . odd.'

'Blink and look away,' said Pug.

Owyn did so and found the lethargy passed. 'What was that?'

'Some magic captures the eye and compels it to follow the cantrip until it's burned into your mind. Let me study this for a while before you try to memorize it.'

Gorath said, 'We had better find some other place to study it. Those snakemen will be coming back here soon.'

Pug rolled up the parchment and looked around the hut. There were other items. Pug didn't have the time to examine them, for Gorath said, 'Too late. Here they come.' He hurried to the back of the tent and cut a long slit down the back. 'This way!'

Pug, Owyn and Gorath hurried through the rear of the tent and fled up a dirt path that led into the hills.

They ran until they found the entrance to a cave. Owyn

used his light spell and said, 'Down there!' He led them deep into the cave and they hid around a turn, listening for the sounds of pursuit.

After a while the silence reassured them the Panath-Tiandn were not following. They sat in the cave. Pug said, 'Well, as long as we're sitting here, hold your hand where I can read this.'

Owyn did as he was told and Pug spent a long time studying the scroll. Minutes dragged by, but Owyn kept the light steady as Pug read.

Gorath grew bored and moved to the mouth of the cave, then out and down the trail a little way to see if there were any signs of pursuit. He returned to the cave and saw that Owyn and Pug were now both lost in studying the scroll.

Knowing there was nothing to do but wait, he set out to explore further along the trail. He worked his way up into a small pass and over a rise saw the trail changing; the stones became smoother, as if once this had been a stone roadway.

With night vision far more acute than any human's he moved effortlessly through the gloom of a night illuminated by alien stars. He hurried along, sensing he was close to something imbued with ancient magic.

He crested another rise and looked down a long trail at a giant cave mouth. Carved into the sides of the mountain were two huge dragons. He halted, torn between returning to get his companions and a desire to explore further. After a moment of conflict, he moved ahead and at a half-trot entered the dark cavern.

James stood, panting, his arms and chest drenched in blood. Six times goblins and moredhel using scaling ladders had threatened to crest the wall, and three of those times he had personally had to beat back the attack. Locklear hurried to his side, nearly dead on his feet and said, 'It doesn't look like they're pulling back for the night! They keep coming!'

'What's the situation?'

'That first tower we stopped with the ballista has been cleared away and now they're moving two of them forward.' Of the half dozen siege towers that were built up north, three had been destroyed by the mangonels on the north wall. Unfortunately they had used all the large rocks capable of disabling the towers and two more had cleared the turn in the road and were pushed up toward the west gate. The first one had been destroyed by the two large ballistas over the gatehouse.

'The ballistas?'

'We still haven't got them repaired. One needs to be completely dismantled and reassembled and the other needs more time to fix than we have. I was thinking if we let them get close enough and then pepper them with fire arrows they might be burning before they reached the walls.'

James looked dubious. 'They can't have neglected to – '

'Hey!' said Patrus, hurtling into view. 'We've got a situation.'

James shook his head to clear it. 'What?'

'Can't you see those bleeding huge towers rolling this way?'

Lacking any humour, James held out a hand dripping with blood and said, 'I've been busy.'

'Oh,' said the old magician. 'Well, there are these two bleeding huge towers rolling up on us.'

'We were just discussing how to fire them,' said Locklear.

'I was saying I can't believe Delekhan's generals could neglect to fireproof them.'

Patrus said, 'I don't know. Why don't we find out?'

He moved past them and lowered his staff over the wall, just as a scaling ladder slammed against the stones. Two fatigued soldiers used forked poles to push it back and from below they heard a scream as a goblin fell from it. Patrus ignored a flight of stones from slingers below that peppered the walls around him.

'It's a good thing they're so bad in co-ordinating their efforts,' observed Locklear. 'If those stones had kept our lads back a moment earlier, those goblins would be over the wall.'

'Give thanks for small favours,' said James.

Patrus aimed the staff at the nearest of the two towers, and spoke a short phrase. A small blast of fire sped from his staff and Locklear said, 'That fireball trick he did in the pass when we met him!'

James turned and saw a third and fourth fireball strike the structure, and could hear shouts from the moredhel and goblins within. Two of the strikes had started fires.

Patrus turned and aimed at the second tower, and missed it with his first blast, then corrected his aim and hit it three times. He managed to generate another half a dozen fireballs at each tower, and soon they were both aflame. Cheers erupted from the exhausted defenders on the wall as a trumpet in the distance sounded.

James sank to the parapet. 'They're sounding retreat,' he said, exhausted beyond imagination. 'We held them.'

Locklear sank to the stones next to him as they heard the retreat from the wall and the defenders took up the cheer. 'What a day.'

Patrus knelt and said, 'Well, that was a good spanking, lads.' The chipper old magician said, 'Don't get too comfortable. There's a lot of work to do before morning.'

Half-dazed from the battle, James said, 'What's in the morning?'

With a cheerful tone that almost caused James to want to throttle the old man, Patrus replied, 'Why, when they attack again.'

Dragging himself to his feet, James said nothing, knowing the old man was correct. He lowered his hand and Locklear reached up, gripped it and pulled himself upright, with a groan of a man four times his age.

Silently they headed down into the keep to begin organizing the survivors for the next assault, while the two siege towers burned like beacons in the night behind them.

Pug and Owyn blinked as they returned to the here and now. 'What?' asked Owyn.

Gorath said, 'You've been studying that scroll for almost a day.'

Pug said, 'It's an alien spell. Very powerful, and it's now burned into my memory.'

Owyn said, 'Mine as well.' He straightened up from his position of leaning over Pug. 'It's a spell to drain energy from those creatures Dhatsavan spoke of, the Rlynn Skrr elemental creatures.'

Pug stood up. 'How long?'

'It's almost dawn.' Gorath pointed out the cave. 'I did some scouting.'

'What did you find?' asked Pug.

'The Valheru place you spoke of. I think your daughter may be there. It is very much like a temple.'

Pug didn't wait and hurried out of the cave mouth. 'Where?'

Gorath followed, then took the lead, showing Pug the way to the cave mouth with the dragons carved on either side. 'A short way inside, steps led down to a huge chamber. I heard sounds like the wind from within and felt an ancient fear, from someplace I cannot name. I thought it best to wait for you two before venturing farther.'

'Wise,' said Pug. 'I think it was very wise.'

Pug, Owyn and Gorath entered and moved down along a stone hallway, and down the stairs Gorath had spoken of. At the base of the steps they entered a huge chamber. Once a host of worshippers could have fit inside, but at present it was empty. At the far side stood two doors of stone. As Gorath had said, a fetid wind seemed to blow through the chamber and it filled all three of them with a terrible dread.

Reaching the doors, Gorath tentatively pushed on one. It was massive, but counterbalanced with great skill, so that as he pushed it swung open slowly, but with ease.

When the opening was wide enough, Gorath released his grip on the door and slipped through, followed by Owyn and Pug. In the next chamber a glowing blue crystal rose from a dais in the centre of the floor, illuminated by a shaft of light from above. Hanging in the middle of the crystal was the form of a young girl, her pale hair floating around her head like a white nimbus.

'Gamina!' shouted Pug.

From out of the gloom on either side of the gem two figures appeared, one from each side of the chamber. They were ten feet tall, the colour of a grey shroud, and their eyes burned like blue ice. Their features were indistinct, shifting and changing, but they appeared powerful in form, with large spreading wings.

Gorath hesitated, but Pug shouted, 'Owyn, the spell!'

Both magicians closed their eyes and for a brief moment Gorath stood, uncertain of what to do. Then he struck with his sword, attempting to slash the creature that advanced upon Pug. His sword passed through the creature as if cutting the air. Only a slight slowing of the blade and a numbing cold shooting down his arm signalled any contact. Then the creature lashed out and sent Gorath flying across the room with a blow that struck like a hurricane.

Then scintillating lights of every colour in the rainbow jumped from Owyn's and Pug's hands, each striking one of the two creatures. The creatures stopped dead in their advance, as if stunned to immobility. The colours whipped through the creatures' bodies, then shot down into the floor, thousands of tiny embers of colour, one after the other. Each bright light seemed to take a tiny particle of the creature with it, and before the two magicians the two elemental beings faded, until at last only a hollow echo of the wind remained in the room.

Gorath stood up and shook off the effects of the blow he had suffered.

Again Pug cried, 'Gamina!' He hurried to the crystal and saw that his daughter was preserved like a living effigy of the goddess of the Panath-Tiandn. He touched the crystal and felt energy flowing through his fingers.

He closed his eyes and traced the patterns of energy in his mind, and at last, said, 'Gorath! Strike here!' He pointed to a facet below the girl's feet.

Gorath didn't hesitate and drew back his sword and with all his might he struck exactly where Pug indicated. The crystal erupted in a shower of gems, splashing the three of them as if a million diamonds had been spilled from a vessel. Pug ignored the falling gems and stepped forward to catch his daughter as she fell. She seemed in a trance, but she lived.

'The gods be praised!' said the magician. His tears flowed as he hugged his daughter to his chest, cradling her as if she were still the little girl who had come to live with his family years before. The mute child who couldn't speak but used her mind like a weapon had become as dear to him as the child of his body. In his heart she was as much his daughter as William was his son.

He gently lifted her chin and whispered her name.

Her eyelashes fluttered and she stirred. '*Daddy?*'

Gorath's eyes widened. He looked at Owyn who nodded. 'I heard it, too.'

She opened her eyes, then she flung her arms around her father's neck. 'Daddy!' She hugged him fiercely and he held her as if he might never let her go. 'He was lying, Daddy. Makala was lying all the time. He tricked me; he gave me something to make me sleepy, and then I woke up here. He said he didn't want to hurt me, but he wanted to get you away from Krondor!'

'I know, sweetheart,' said Pug softly. 'It's all right. We're going home now.'

'How?' asked Gorath.

Owyn said, 'There's supposed to be a gate here, according to Dhatsavan. By that I suspect he meant a rift of some sort.'

Pug looked around the chamber and said, 'I see nothing here.' He turned to his daughter and asked, 'How do you feel?'

She stood and assured him, 'I'm all right, really.'

Owyn stared at the girl, barely into her teens, and was struck by what a beautiful woman she would become. She caught him staring and smiled and he turned away, blushing.

Pug smiled and said, 'You remember Owyn and Gorath from Krondor, I presume?'

'Yes,' she said with a shy smile. 'Thank you for helping my father find me.'

Gorath said, 'It is our honour.'

Owyn just smiled and nodded.

They moved across the chamber and found another corridor on the other side of the hall. Another large door loomed before them, and Gorath opened it. It led them into a chamber in which a huge wooden device stood.

Pug took one look at it and said, 'It's a rift machine!'

'Are you certain?' asked Owyn.

'I did more research on rifts on Kelewan than any other Black Robe,' said Pug. 'But even if I hadn't I'd have recognized that device. It's Tsurani.'

'Can we use it?' asked Owyn.

Pug went to it and examined it for a long time, then said, 'It's been deactivated.'

Owyn said, 'Deactivated?'

Pug grimaced. 'Turned off. It's not working.'

Gorath said, 'You mean we're stuck here?'

Pug sat down and said, 'Unless I can come up with a way to get it working again, yes, we are stuck on this blasted world with no way to get home.'

Gamina put her arms around her father's neck and Gorath and Owyn both sat down on the floor, for want of anything better to do.

EIGHTEEN

Regroup

Smoke blinded the defenders.

James had managed to sleep for an hour, Locklear for two, and as ordered, soldiers had stood watch on the wall through the night, sleeping in shifts.

James now squinted through the smoke, from his headquarter position on the gatehouse, as the smouldering ruins of the two siege towers filled the air with an acrid haze. Even the morning breeze wouldn't help, since it would continue to blow the smoke at the wall. The night sky had lightened as the sun rose behind the defenders. Soon it would clear the top of the mountains. Some time between now and then, James knew, the enemy would attack again.

He looked down and saw bodies floating in the moat, both attackers and defenders. They looked thick enough to walk over to reach the drawbridge, he thought.

Reports had been coming in from every position of defence in the castle and James knew the sickening truth: they could not hold another day. Unless the attackers were criminally stupid or fate took a hand, the castle of Northwarden would fall before sundown.

James had already conceived half a dozen ways he could take the castle were he commanding the attackers, then had tried to imagine countering each of those offensives. Each time he came away realizing he just didn't have enough men if they tried anything other than a single-front assault. Something as basic as storming the gate road while sending goblin climbers up the north slope once more would overtax his defenders and make it impossible to stop one of the two fronts.

Locklear came and asked, 'What do we do?'

James said, 'I'm thinking of abandoning the outer wall and moving all the soldiers into the inner keep.'

Locklear shook his head in an exhausted admission of defeat. 'I can't think of anything better to do. It will make them spend more lives and waste more time taking the castle.'

'But it will make it impossible for us to hold.'

'Do you think we have any hope of holding?'

'Right now I'm trying to come up with a brilliant plan to sneak around behind Delekhan's lines and attack him from the rear.'

A sergeant, still covered in blood from the day before, approached. 'Report,' said James.

'Three more men died during the night, squire. We have one hundred and fifty able-bodied men on the walls, another seventy walking wounded who can still fight, and some of the more mobile injured are helping out in the Great Hall.' The Great Hall had been converted to an infirmary where nearly a hundred soldiers of Northwarden lay dying for lack of the skills of a healer.

James shook his head. 'Let the men rest until the enemy attacks again. Get as much food and water to the men on the walls as you can. The only way we get another hot meal is to win this battle.'

The sergeant said, 'Yes, squire,' and hurried off.

Patrus came walking up the steps that led up the wall to the gatehouse, looking very tired. The old magician said, 'I've done all I can with the wounded. What can I do here?'

James said, 'Figure out a way to keep the enemy away from one of two places, the north wall or the east gate; either one, I don't care.'

'Too much wall and not enough soldiers?' asked the old man.

'Something like that,' said Locklear.

Patrus said, 'Well, if they don't clear away all those bodies down there on the road before they attack again, I can help you out on that front. The more metal down there touching the ground, the better. Move some of your boys to the north wall.'

'What can you do?' asked James.

With an evil grin, the old man said, 'What, and spoil the surprise? No, you just wait, sonny, and when the time comes, I'll give you a show.'

'I'm not interested in a show. How much time can you buy us?' asked James.

'A few hours, depending on how much courage those moss troopers can muster after I smack them around a bit.'

'Give me two hours to defend the north wall before I have to turn my attention to the east gate, and we may buy ourselves another day.'

'You just watch me,' said Patrus. 'Now, I've got to go to my room and get a few things.' He hurried off.

Locklear turned to James and, despite his exhaustion, said, 'Isn't he about the most evil old man you've ever met?'

'No,' said James. Then he smiled and added, 'But he does get close.' Drums sounded in the distance, and James announced, 'They're on their way.'

Shouts from the north wall alerted James that goblin climbers were again trying to work their way up the face of the cliff. They had exhausted their supply of stones to scrape the climbers off the cliff, as well as every piece of furniture, crockery, kitchen utensil and tool they could spare, and most of the water they didn't need for drinking had been boiled and spilled. Now they were forced to spend valuable arrows trying to pick them off one at a time, exposing their own archers to fire from below.

Patrus returned and said, 'Give me room.' He sat down on the stones, crosslegged, and put a small bowl in front of him. 'It's taken me a week to get everything ready for this. Now, shut up and don't disturb me unless the world's about to end.'

He dumped the contents of a small pouch – a lumpy mass of powders and what seemed to be small stones or rocks – into the bowl then closed his eyes. He chanted a short phrase, opened his eyes, and extended his index finger. A small flame erupted from the end of it, and he lit the contents of the bowl. Instantly the flame transferred to the contents of the bowl. A green and blue cloud of smoke, far thicker and more abundant than either James or Locklear would have thought possible, billowed up out of the bowl, and reached the stone ceiling of the gatehouse. The smoke seemed to recoil from the stones and Patrus waved his hand over his head, palm toward the eastern road, as if blowing the smoke in that direction.

Obedient to his gesture, the smoke rolled out the front windows of the gatehouse, thinning as it expanded, and looking more and more like clouds as it hung above the road. James looked and saw a tightly-packed formation of hide-covered shields in the van, a company of goblins marching with trolls behind them. The apelike trolls had massive shoulders on which they easily carried scaling-ladders, and each had a shield on the outside arm, with a warhammer or axe dangling by a leather thong.

'Troll assault troops?' Locklear asked.

'So it seems,' said James. 'I've not heard of any such before, but if they're serious about coming up those ladders, we have a problem.' Trolls were not significantly better fighters than goblins or moredhel, but they were a great deal more difficult to kill. Whoever led the opposing forces must have correctly guessed that the defenders were bordering on exhaustion.

In the smoke from torches and the smouldering towers, Patrus's mystic smoke was hardly noticed. As they watched, James and Locklear both saw that the smoke was thicker.

As the attackers came within bow-range, archers on the walls began firing. James was appalled by how few arrows were flying from the defenders. He could taste defeat.

Then a low rumbling started below the castle and James touched the wall. He felt the low thrum of energy coming from the earth.

The attackers took no notice of it until the level of the vibration became obvious to marching feet, even to those running forward with the heavy ladders. The attack faltered.

Then Patrus let out a cackle and shouted, 'Hang on, boys!'

The castle seemed to heave.

A full half of the attackers were knocked off their feet. The sound of the earthquake drowned out the noise of battle.

And then the sky exploded.

A bolt of lightning struck the armour of an attacker on the ground, knocking down a full dozen comrades around him. It was followed the barest instant later by an explosion of thunder, which made the ears ring. The air reeked of the acrid lightning smell and the stench of burning flesh. Moredhel, goblins and trolls lay writhing in agony, their skin smoking from the flash.

Then another bolt struck the ground a few feet away, killing another dozen. An instant later, a bolt struck a moredhel with an upraised sword, illuminating him in a blinding white flash for an instant before he exploded in a fireball, killing most of those standing near him.

James ducked behind the wall, and yanked Locklear by the tunic, pulling him down. 'Get behind the wall!' he shouted to the men atop the gatehouse, and the order was relayed along the eastern wall. Bolt after bolt erupted from Patrus's mystic cloud and each was accompanied by a monstrous peal of thunder. Men clutched their ears lest they grow deaf from the sound of them.

James wished he could somehow crawl down the stairs

and reach the haven offered by the lowest basement of the castle, then wondered if that would be deep enough. He could barely imagine what it was like for those exposed on the road below.

Over and over the lightning blasted, until suddenly there was silence. The instant the noise stopped, the vibration of the earthquake ceased as well.

James leaped up and looked over the wall to see the army that had only minutes before been attacking now in total rout as it fled down the hill. At least a thousand attackers lay dead on the road leading to the castle, many trampled to death by their own comrades.

James knelt down next to Patrus who blinked his eyes and said, 'How'd that do?'

'It did the trick. They're in total flight.'

Locklear leaned over behind his friend. 'What do you call that?'

'Don't have a proper name. It was taught me by a fellow down in Salador, who had learned it from a Priest of Killian, but he had to change it. I think of it as "Killian's Rage".' He stood up. 'Always wanted to try it out, but never had anyone I was mad enough at to risk it.' He moved to the wall and looked between two merlons. Noting the number of bodies, he said, 'Worked better than I thought.'

James shouted, 'How's the north wall?'

A voice called back, 'They fell off with the earthquake.'

James put his hand on Patrus's shoulder. 'You bought us some time.'

Locklear sank down next to where they stood and leaned back against the stones. 'I can't move.'

James reached down and hauled him back to his feet. 'You must. They will be back. Unless Patrus can duplicate that little surprise again?'

Patrus shook his head. 'If I had the makin's, but it would take a while to put it together, and I'd have to get out in the woods and look around a bit.'

Locklear said, 'One thing bothers me.'

'What?' asked James.

'Where are their magicians?'

James's eyes widened. 'Gods! If that little display didn't bring them running, they're nowhere near here.'

'What's that mean?' asked Locklear.

'It means we've been duped.'

'I don't understand,' said Locklear, sinking back down to sit on the stones.

'If they're not here, they're somewhere else!' said Patrus. 'I know you're tired, but that's no excuse for stupidity!'

'Leave me alone,' said Locklear in feigned self-pity. 'I'm enjoying my delusion. Even now I just imagined I heard a Kingdom trumpet blowing in the distance.'

James halted, and listened. 'You're not deluded. I hear it too.'

James climbed up on the wall, his youth as a thief giving him the keen balance and steel nerves needed to step atop the merlons of the wall and stare into the distance. The smoke was still making it difficult to see, but after a moment, James shouted, 'I see Arutha's banner!' He jumped down, and said, 'Lower the drawbridge!'

James hurried down the steps, rejuvenated, with Locklear and Patrus following. By the time they reached the marshalling yard, the portcullis had risen high enough for James to duck under. He did so and ran to the end of the lowering drawbridge, jumping off before it touched ground.

He had his sword in hand in case one of the bodies wasn't as dead as it looked, but by the time he reached the bottom of the road, Arutha and his personal guards were riding to meet him. Stopping before the monarch of the Western Realm, he said, 'I was beginning to believe you were going to miss all the fun!'

'I wouldn't have you think I was impolite,' said Arutha. 'How are the men?'

'Doing badly. Baron Gabot and his officers were murdered. Most of the men are dead or wounded, the few that aren't are exhausted. Another day and you would have found us all dead. Not to sound impolite, but what took you so long?'

'We came as soon as we got word. Your messengers were ambushed and abducted and it took them a little while to escape and reach me. They arrived only three weeks ago. What of support from the south?'

'None. I sent word to Romney, Dolth, even to Rillanon.'

'Others may be coming,' said Arutha, 'or those messengers were also ambushed. Owyn told me that you killed the head of the Nighthawks, but they still must have had agents in place before that.'

James said, 'I fear we may never truly see that nest of murderers obliterated. They are like the legendary snake of the Keshian Underworld: cut off its head and it grows back.

'But more to the point, we think all this may have been a ploy.'

Looking at the evidence of destruction all around him, Arutha said, 'An expensive ploy.'

'But a ploy nonetheless. When Patrus, the old magician we met up here, used his magic, there was no answering magic.'

Arutha said, 'What of those who are called the Six?'

'When we last heard they were still in the west.'

'The west!' Arutha swore. 'This may have been a terrible ploy, one sold convincingly on the lives of thousands of soldiers to get us to move from the Dimwood.'

'Have you moved all your forces?'

'No, the garrison near Sethanon was left in place, but I brought the rest of my companies with me. I will send patrols up into the passes to see how many of the enemy are arrayed against us.'

Arutha looked worried, an expression James had seen on many occasions, and rarely without justification. 'Let's get to the castle, relieve your command, Seigneur, and sort this out.'

Arutha turned and passed orders to a young officer, then said, 'I've left Gardan near Highcastle, and Captain Philip at the Sethanon garrison. I think between the two of them we can hope they'll keep Delekhan from achieving an easy strike by this ploy.' Then he looked at James. 'But after you've eaten, slept, and eaten again, I want you and Locklear and a fast patrol heading back to Sethanon.'

James grimaced. 'Those long rides, again?'

'I've got a healing priest with us; I'll ask if he has anything to ease your pain.'

James looked at Arutha to see if he was joking and when James saw he wasn't, said, 'Very well.'

Arutha's concern was obvious and James asked, 'What is it, Highness? I've known you too long not to recognize that look.'

'Just worried about Owyn and Gorath. I sent them to fetch Pug because of what they said about the Six, but if they were ambushed between Malac's Cross and Krondor, or if Pug had left Krondor on one of his mysterious jaunts and Katala couldn't find him, or any number of such problems, well, when the Six appear, I suspect we would be well served by a magician.'

James grinned. 'I have one.'

'Someone responsible for that lightning display we witnessed as we approached?' Arutha mounted his horse.

'Yes.' James started walking back toward the castle. 'He's an original and I think you'll find him entertaining. At least for a few minutes.'

Arutha smiled his half-smile and James felt better for that.

Gorath looked at the snare and kept motionless. The creature looked like an armoured rabbit, or a turtle with long legs, but either way it was the only edible creature they'd encountered so far that wasn't an insect. Two other creatures had proven inedible after being caught. This world abounded in insects, from tiny gnatlike fliers that would swarm to plague Gorath when he tried to remain motionless, to cockroachlike creatures that were as long as his forearm.

He had identified half a dozen edible roots and a prickly fruit that tasted like a sour melon and possessed a tough, stringy pulp, but which contained a lot of water.

They had found a well near the ancient temple, and had created a water bag out of an old piece of leather they had found in the temple.

Gorath!

Just a moment, he tried to think back. It was still difficult for Gamina and him to speak compared to her and the other humans, but he was getting better. He had to focus his thoughts. He imagined himself shouting at her. *I am about to catch supper,* he thought.

He received a non-verbal sense of patience.

The armoured rabbit moved and he pulled the snare, tangling the right hind leg. He was on the creature in a moment, and having learned by trial and error, had the creature upside down, so it was forced to stick its neck out. He broke it and quickly had it out of the shell. He had learned to their collective distress that if you didn't get the creature out of its shell within minutes of killing it, the flesh quickly tainted and the resulting stomach distress was extremely unpleasant. He cut the meat out of the shell and deposited it in his travel pouch.

He turned and hurried toward Gamina. *What is it?* he asked, knowing she'd hear his mind before her ears would register his words.

Owyn and Father have found another cache of mana.

'Do they think we have enough?'

Maybe, she said, as he hiked into view.

She turned and he followed her down the path to the

entrance to the abandoned temple. For whatever reason, religious prohibition, fear of the Valheru, or fear of Gorath, Pug and Owyn, the serpent-men had not attempted to enter this area.

They had attacked the second time Pug and Owyn went looking for more mana, for Pug had a plan to activate the abandoned rift machine. Gamina had tried to read their minds and had come away confused, for the Panath-Tiandn, who called themselves the Shangri, were a strange mix of very simple and very clever thinking. They were primitive and superstitious in their daily life, almost animalistic in their thinking, but brilliant in the manipulation of magic. Pug commented that it was ironic they were imprisoned on a planet, which they called Timiri, where magic had to be harvested like a crop.

Pug had declared them magic artisans, probably responsible for the construction of devices for Alma-Lodaka. Given his experiences with the Pantathians, who were obviously related to the Panath-Tiandn, Pug assumed that the ancient Valheru had intentionally limited the scope of their intelligence, keeping it focused where it served her.

How they had managed to survive on this blasted world was rapidly becoming apparent to Gorath and the humans, for they had run out of food two days after finding Gamina. It had been a week since, and they were attempting to gather enough of the crystal magic for a plan of Pug's. Gorath was unsure of how these pieces of 'frozen magic' would serve, but he was content to let the human magic-users struggle with that problem. He had elected to concentrate his attentions on finding food. Like many places that are apparently barren at first glance, this world was teeming with life if you knew where to look for it.

Since discovering the rift machine, they had explored the entire island, save for the peaks above the temple. The island was large enough that it took Gorath three days to travel from the northernmost point, where the seven pillars of the gods were located, to the southernmost point. It was roughly half that time to travel east to west, though the journey couldn't be conducted in a straight line due to the rise of mountains down the centre of the island.

They thought there might be land to the west, or at least Pug thought it likely, making the observation after watching the sunset one night. He had mentioned the effects of light over the water and thickness of clouds and other factors which were only interesting in the abstract, at least to Gorath. Unless they needed to travel to that distant land to find more of the solid magic.

Gamina had a fire ready when Gorath reached the cave and put down his kill. 'Are we going to try to put your father's plan to work tonight?'

'I don't know,' she replied.

Gorath watched her, and was forced to admit she was an admirable child, even by his own people's standards. He knew little of human children, but knew she had to have been subjected to a frightening experience, yet she was calm, focused, and relatively cheerful considering the circumstances.

She was also quite beautiful, after the human fashion, if Gorath could judge such things. She certainly seemed to have Owyn's attention, though Gorath could tell he was being circumspect either because of her father's presence or her youth. Perhaps Owyn dreamed of years to come. Again, Gorath was uncertain of these human conventions.

Owyn and Pug appeared with a large bundle of cloth, one of the woven doors pulled from a hut. Pug had observed that with so many huts and so few inhabitants, the population of this area must be falling. He had wondered what the rest of the planet looked like, but had been unwilling to use any of his arts to explore, fearing they needed to hoard as much of this solid magic as possible.

'I think that should do it,' said Pug as they put down the bundle.

'Good,' said Gorath. 'I tire of these creatures as our only catch. I would even welcome those stale breadcakes we ate in the mountains, Owyn, for the change they would bring.'

'As would I,' said the young magician.

'What do we do if this doesn't work?' asked Gorath.

Pug said, 'Then we explore the rest of this island, and if there is no way to be found here, we do what we must to build a boat and make our way westward, to the next body of land.'

Owyn closed his eyes and put his thumb and finger to the bridge of his nose.

'The headaches, again?' asked Gamina.

Owyn said, 'Yes, but it is passing.' Owyn had been experiencing intermittent but severe headaches since having shared the Cup of Rlynn Skrr with Pug. 'And it hurts less than before.'

Pug said, 'When we return to Midkemia, I think, my young friend, you will discover you have powers you never anticipated.'

Sighing, Owyn said, '*If* we get back.'

Pug looked at Owyn and his expression was without doubt. 'We will get back.'

Owyn said, 'Very well. What else do we need?'

Pug said, 'Nothing but knowledge.' He asked Gorath, 'Have we explored every chamber in this complex?'

'Yes,' said Gorath. 'As I told you.'

Pug said, 'Then we should plan on attempting to return tomorrow.'

'Why not now?' asked Gorath.

Pug said, 'Owyn and I will need as much rest as we can before we attempt this. I know much about rifts and their nature, but that machine is of alien design and may not work as the Tsurani machines I'm familiar with. Therefore, I would not care to make a mistake because I was tired. So, in the morning, after we sleep, then we shall try.'

Gorath nodded.

Owyn lay back, tired from the long walk carrying the mana. 'Gorath, can I ask you a question?'

'Yes, Owyn,' said the dark elf.

'When you bowed before the Queen, I take it that was some ritual, but I don't understand it.'

Gorath sat back on his heels, thinking. At last he said, 'When I first beheld Elvandar, I called it Barmalindar, the name of the legendary world of golden perfection all elvenkind believes is its ancestral home.'

'Fascinating,' said Pug. 'I have spoken to Prince Calin and Tomas and other elves, but this is the first I had heard of such legends. I assumed you were originally from Midkemia.'

'We are, as were the dragons and the Valheru, but there is a spiritual source to our race, beyond Midkemia,' said Gorath. 'When we die, we travel to a Blessed Isle, where we join with the Mothers and Fathers who have gone before. But we all come from Barmalindar.'

Gorath looked at Owyn. 'From time to time, one among my people will hear a call, a tugging, that will compel him to travel to Elvandar. My people will hunt such a one down as a traitor if they can, and kill him before letting him reach Elvandar.' Gorath closed his eyes and his tone was tinged with regret. 'In ages past, I did so. But a few get there and those of the eledhel call them "returned". They take new names and it is as if they had been eledhel all their lives.'

'What I don't understand,' said Owyn, 'is the Queen saying you had not finished returning. What does that mean?'

'I still have ties to my past, an obligation which prevents me from completely joining my kin in Elvandar.'

Owyn asked, 'What obligation? I thought your children dead and your wife had left you?'

Gorath looked at Owyn, and said, 'I must kill Delekhan.'

Owyn said, 'Oh,' and lay back against the cave wall. They all remained silent while Gamina cooked and Pug prepared for his attempt to revive the rift machine the next day.

*   *   *

James had witnessed torture before, but he took no pleasure in it. Yet Arutha was desperate to learn Delekhan's plan.

The prisoner was some sort of chieftain or captain, but someone who was obviously in a position to know more than the common trolls and goblins who made up the bulk of this company. The half a dozen renegade humans who had been captured made it clear this moredhel was the only one who might know what was going on.

And Arutha knew something profoundly disturbing was going on.

They had sent scouts up the pass and discovered there was no second force waiting to support the first. The force that had been broken at Northwarden was the total of Delekhan's army in the area. Thousands of warriors, goblins, trolls and the magicians known as the Six were somewhere else.

The moredhel groaned as the ropes were pulled taut. His feet had been tied to two iron rings in the floor, two ropes had been tied to his wrists, and those ropes thrown over a ceiling beam, making a makeshift rack.

Arutha spoke in even tones. 'Speak, and you'll see your children grow to adulthood, moredhel. My word on it. I'll turn you loose as soon as you tell me what I need to know. Where is Delekhan?'

The moredhel looked up, and instead of fear or even hatred, James saw amusement in his face. 'What does it matter, Prince of the West? If I tell you, you cannot prevent my master from reaching his goal. Release me from these ropes and I will tell you exactly where Delekhan is.'

Arutha nodded and the ropes were released, letting the moredhel fall to the stone floor. Looking up with a glare, the moredhel spat, 'Delekhan rests in Sar-Sargoth and gathers his army there.'

A captain of the Royal Guard made as if to strike the

moredhel, saying, 'Lying dog,' but Arutha gripped his hand, preventing him.

'Why would your master sit on the throne of ancients, while you and your companions spill your blood here in Northwarden?'

'Because you are here, now, Prince,' said the moredhel.

'But I have an army at Highcastle, and another at the Inclindel Gap.'

'It does not matter, Arutha. Only one small garrison have you left in the Dimwood, and within days my master shall overrun it and the prize shall be ours.'

Arutha's eyes narrowed. 'Days . . . ?' He stood up. 'Gods! They're going to use a rift!'

James demanded, 'How is that possible?'

Arutha said to the captain, 'Take this one to the gate and turn him loose. I'll not forswear, but give him no weapon, food or water. Let him plunder his dead comrades if he wishes to survive.'

Soldiers roughly picked up the injured moredhel and half-dragged him out the door. James asked, 'Highness, how can they have a rift?'

'The Tsurani know how to make them, as does Pug. And we have suspected the Pantathians knew of their making,' Arutha said. 'Whatever the source, if Delekhan can fashion one, that moredhel chieftain is right. It is then but a short step from Sar-Sargoth to the Dimwood and I am in the wrong place.'

'What should we do?' asked James. He was still tired, but had spent a restful night after drinking a special herb tea made for him by Arutha's healer.

Arutha said, 'I shall have a company of gallopers accompany you, that magician character and Locklear to the Dimwood. Kill the horses if you must, but ride until they drop. I'll have Father Barner make up restoratives so you don't kill yourselves as well. I'm sending you first to Highcastle. Tell Baron Baldwin to strip the garrison and march on Sethanon.

Then pick up Gardan's company and get to the Dimwood as fast as you can. I will be following as fast as I can turn this army around and get it moving.

'But you and your two companions must ride to the Dimwood, even if you have to leave the soldiers behind. Find Captain Philip and tell him to start looking for that rift machine. If it can be destroyed before Delekhan can bring in the bulk of his army, we may still stop him.'

'He could be there already,' said James.

'Which means you can't start any too soon,' said Arutha. 'Get mounted and leave now. You've got half a day left.'

James bowed and hurried to find Locklear and Patrus. He knew neither of them would be happy to get these orders. He knew he wasn't.

NINETEEN

Encounter

Pug motioned the others to stand back.

He piled the physical mana up, then took his and Owyn's two crystal staves. 'As I see it, this gate ceased working when the connection on the other side was broken, for that must have been where the power originated.'

'But where is the other side?' asked Owyn.

'Somewhere in the palace at Krondor, or nearby. They could have rendered Gamina senseless and taken her anywhere, and then to this world, but the trick Makala pulled with creating that temporary portal in my study required the original gate to be very close.'

Owyn asked, 'Why?'

'When we have time, I'll explain the theory behind it, but leave it for the moment that the device Makala gave me didn't have the power to transport me between worlds. It simply keyed me into a rift that existed nearby.'

Owyn seemed to understand. 'Did you ever discover why Dhatsavan and the other gods of this world froze the mana like this?'

Pug shook his head as he stood up and backed away from the machine. 'I think it was a desperate ploy; if they did this at the height of the battle perhaps the Valheru judged the world destroyed and moved on; they might have sensed all the magic going away and feared they would be trapped here. We may never know exactly why. And I'm in no mood to return to the pillars to discover why.'

'How do we calibrate this thing without a power source?' asked Owyn.

Pug held up the Cup of Rlynn Skrr. 'This is what we will channel the power with. Dhatsavan told you it was a key, and with it you could reach other worlds.'

'Of course,' said Owyn.

Pug pointed to his head. 'Here I have the knowledge. And for that I'll need your help.'

Owyn blinked. 'How?'

Pug said, 'I need to warn all of you: this is something that may not work, or might even lead to our destruction.' He addressed the last to Gorath and his daughter. 'I wish I could spare you the risk, but my experience with rifts tells me we shall have but a few seconds to attempt this gate.'

Gamina said, 'Just tell us what to do, Father.'

'After I instruct Owyn, I will count aloud, and when you hear me say "three" we will attempt to activate this rift gate. If this works, a shimmering silver light should appear between the two poles and turn an opaque grey. The instant you see it turn grey, jump between the poles. It might help if you and Gorath held hands. Owyn and I will follow a moment later.'

He showed them where to stand. Then he said to Owyn, 'This is perhaps the most difficult thing you have ever been asked to do. You have both the energy and knowledge, though the second is only recently within you.'

Owyn said, 'I don't understand.'

Pug nodded. 'My powers are still weakened, because of my misuse of the cup. While my memory has returned, it may be some time, days perhaps, before I can recapture even part of my powers. But you have power, what you brought with you, as well as knowledge gained from me when we shared the cup.'

'But I don't know anything about rifts,' Owyn objected.

'Close your eyes and stop trying to think about what to do. Just relax and let things come to you.'

Owyn looked dubious but he closed his eyes. Pug waited a few minutes and said, 'You're still trying. Think of something else?'

'What?'

'Think of something boring, perhaps a tome you read a long time ago that didn't interest you, or one of Elgohar's lectures at Stardock.'

Owyn laughed, and suddenly something entered his mind. 'I . . . wait, something . . .' He opened his eyes. 'I laughed and there, for a moment, I knew something about rifts.'

'Close your eyes and think of something else,' said Pug.

Owyn took a deep breath and closed his eyes again. He let his mind drift and memories came flooding back. He remembered his older brothers mocking him as he studied the few books his father possessed, and he remembered a girl in the town he liked but never spoke to, though she probably would have been flattered by the attention of the Baron's son. Then he remembered meeting Patrus and that old scoundrel's less than deferential attitude to his father, and his deep wisdom buried under that rough, country-bred exterior. He could see the old man in his mind's eye and could even hear him talking about the spells he could teach the boy.

'I tell you, boy,' said the memory of Patrus, 'the trick to making fire erupt from your fingers is nothing. You just have to want the air around the finger to get hot, and if you get it hot enough, if you want it enough, then the fire appears.' Owyn remembered trying that skill for hours until suddenly it happened.

Then in Owyn's memory, Patrus was saying, 'The structure of a rift is best understood if you ignore common references to three-dimensional location. The idea of being "here" and then being "there" is an impediment rather than a help. If you think of a rift as a "gate" between two places that can be side by side at any moment, if you will . . .' Patrus's voice droned on in Owyn's mind, but the lesson was so obvious as Owyn listened.

Suddenly Owyn's eyes opened. 'I know what to do!'

'Good,' said Pug. 'I once had Macros guide me while I utilized my power to enter and destroy a rift, so now I'll guide you. Gamina?'

'Yes, Father?'

'If you could link Owyn's mind with mine, and then get ready to jump, please.'

She took Gorath's hand and said, 'When you're ready.'

Pug nodded and Owyn suddenly felt their minds linked by Gamina's. Then Pug held out the Cup of Rlynn Skrr in one hand, and the crystal staff in the other. He put the staff firmly on the ground and said, 'Ready?'

Owyn put his staff into the pile of mana and said, 'Ready!'

'One, two, three!'

Owyn took the energies given him by the staff in his hand and let them flow down the staff to the pile of mana that lay at the base of the rift machine. He willed it all to move back up the staff, through him, to Pug to the cup he held. A blinding flash of energy erupted from the cup and filled the space between the poles.

It was like reaching out and moving aside a curtain, thought Owyn. He even knew where he wanted the rift to form. He opened his eyes and saw a shimmering silver wall turn grey before him, then he saw Gamina and Gorath jump and vanish through it. 'Now!' shouted Pug, and Owyn, still gripping Pug's hand, jumped after his companions.

He felt himself pass through a grey void, a moment of sensationlessness, and then they were stumbling on a stone floor, trying to keep from falling over Gorath and Gamina.

The room was dark, with a faint light entering the window.

'Where are we?' asked Gorath.

Pug laughed. 'In my study, in Krondor.'

Gamina jumped to her feet, clapping her hands in delight. Even before she could speak, the door flew open and Katala hurried in, wearing her nightdress. She clasped her daughter to her heart then put her arm around her husband. 'I knew you'd find her,' said Katala.

Pug luxuriated for a moment in the presence of his wife and child, then said, 'Makala has much to answer for.'

Katala said, 'The Tsurani? He's behind your disappearance?'

'It's a long story, my love. Gamina will tell it to you when you're both safely at Stardock.'

Katala looked at her husband and said, 'Where will you be, husband?'

Pug looked at Gorath and Owyn. 'We must go to Sethanon.'

James looked at Patrus and the old magician shouted, 'Worry about yourself, boy. I can take care of myself!'

James was inclined to agree. Arutha's healing priest had concocted a restorative. No matter how tired they were when they lay down to sleep, by sipping a tea made from the magic herbs, they were fully restored the next morning.

They had ridden like madmen, running horses into the ground, trading mounts three or four times a day, commandeering mounts along the way. They had warned Baron Highcastle and allowed themselves one night of luxury, sleeping in a bed, and they left the next day with a new retinue and remounts, while the exhausted soldiers who hadn't had the luxury of the magic restorative would follow after with Highcastle's garrison as it force marched to the Dimwood.

They saw tents and banners in the distance, as they crested a rise in the road, and headed toward the northern boundary of the Dimwood. They slowed as Kingdom sentries flagged them down. They reined in before a sergeant of the Royal Krondorian garrison, who recognized both squires. 'Where's the Knight-Marshal?' asked James.

'In the command tent with the Duke, sir.'

'Which duke?'

'The Duke of Crydee, squire,' said the sergeant.

James asked, 'Martin's here? Good.' He motioned to the others to follow and headed toward the large pavilion tent which dominated the encampment.

Reaching it, James reined in and dismounted, handing the reins of his horse to a lackey. The others were a step behind him, though Patrus looked as if the long ride was finally wearing him down.

At the entrance to the pavilion, James said, 'Tell the Knight-Marshal Squires James and Locklear are here.'

The guard disappeared inside, and a moment later motioned for them to enter. James and his companions went into the command pavilion and found Knight-Marshal Gardan studying the map on the table before him. Looking up, his dark, wrinkled face split into a grin, one that seemed to light up the tent. Yet his eyes looked troubled. 'I hope your arrival means Arutha is coming right behind.'

James said, 'He's on his way, as is Highcastle, and should be less than a week behind.'

'A week!' said Gardan.

'We ruined some horses getting here,' said Locklear.

'You're to move at once to the Sethanon garrison,' said James. He looked around. 'I was told Martin was here.'

'He was,' said Gardan. 'But he and his trackers are already heading into the woods.'

'What brought him here?' asked Locklear.

'Tomas sent word to Crydee and suggested it might be a good idea for some special talents to appear here in the woods.' He pointed south. 'Martin, his trackers, and a company of elves are out there skulking through the woods, looking for Dark Brothers.'

'So Delekhan's on his way?' asked James.

'He's here,' said Gardan. His finger pointed out a point on the map, a 'V' formed by two rivers that ran through the

Dimwood. 'He showed up behind us two days ago. I don't know how he did it, but instead of being in front of us, he was behind us.'

'We think he's using a rift machine,' said James.

'Damn,' said Gardan. 'If the Riftwar taught us anything it was that once one of those damn gates is open, they can walk soldiers through as fast as they can get them lined up and on the march.'

'Can't be a big one, though,' said Patrus.

'Who's this?' asked Gardan.

'Patrus is my name,' answered the old magician. 'If it was big, he'd be on the march.' Patrus pointed to a place to the southwest of where Gardan had pointed. 'Is there a ford or bridge around there?'

'We're trying to find out,' answered the Knight-Marshal.

'Come on, boys,' said Patrus.

'Wait!' said James. 'Where are we going?'

'If that pointy-eared devil is already here, then he's getting ready to move out. If he's not here yet, and we can close that damn machine down before he comes through, his boys'll start running north like their tails was on fire, if they had tails.' He grinned. 'Those dark elves just don't like being far from home.'

Gardan looked at James. 'Who *is* he?'

'It's a long story,' said James. 'I'll tell you after all this is over. But he's the only magician we've got, and unless Pug shows up we've got to make do.'

Patrus made a face. 'Make do? I ought to go home and let you sort this out.'

'Sorry,' said James. 'Tired, that's all.'

'That's all right, Jimmy. You get me to that damned machine and I'll shut it down.'

Locklear looked sceptical. 'I was under the impression that it was a special sort of magic, rifts, I mean.'

Patrus said, 'Special to make, no doubt. What little I've heard about rifts isn't good; most of the time they show up because a magician made a mistake.' He winked. 'But messing magic up, that's easy.'

James said, 'I hope so, because if we can cut off most of Delekhan's army before they get here, we just might keep the lid on this boiling kettle until Arutha and the rest of the army get here.'

Gardan made a sweeping motion with his hands. 'We have a very fluid front. They're dug in here, here and here,' he said, pointing at three different places along a river. 'Lord Martin went this way.' He indicated a pass between the two closest points. 'He thought he might be able to slip past the dark elves.'

James shook his head. 'If there's a human who might slip through the woods past dark elves, it would be Martin Longbow, but even that's a little hard for me to imagine.'

Gardan smiled. 'We're providing him with some distractions.'

James returned the grin. 'Well, then, if it's loud and lively, we might be able to follow after him.'

Locklear said, 'Are we ever going to get any rest?'

Patrus looked at the younger man with an expression of contempt. 'Get killed, boy, and you'll get all the rest you'll ever need. Now, come on and let's get after it.'

Locklear nodded in resignation. James said, 'We have a better chance if we travel light. Do you have any scouts still in camp?'

Gardan shook his head. 'No, they're working with Martin and the elves along that river. If you're lucky, you'll bump into them before you meet any of Delekhan's forces.'

James said, 'Well, we've got a good half-day's light left, so we should probably head out.'

'How are you fixed for stores?' asked the Knight-Marshal. 'We have enough food for a week,' said Locklear. 'We loaded up before we left the last change of horses.'

Gardan said, 'Then let me give you the current password,

"Krondor's Eagle", as we have some human renegades from Queg mixed in around here, too.'

Locklear said, 'After I got Gorath to Krondor, I was hoping I'd never see another bloody Quegan mercenary again.'

Gardan said, 'You have my permission to kill as many as you like when you meet them, squire: now get out of here.'

James laughed and led Patrus and Locklear out of the tent. He pointed to their horses and asked a guard, 'Any fresh remounts nearby?'

The guard said, 'Over there, squire. Captain Philip's taking care of cavalry for the Marshal. He can set you up.'

They took their horses and led them toward the cavalry command, and Patrus said, 'Oh, joy! Another horse to ride!'

Locklear said, 'Would you rather walk?'

'Right now, boy, you bet I would,' was his answer, followed by his signature nasty laugh.

James shook his head. He was trying to like the crusty old magician, but he was finding that as time wore on, it was getting harder to find reasons not to strangle him.

James motioned the others forward, and they rode within sight of a recent skirmish. Three renegade humans and one moredhel lay dead in a clearing, and from the number of arrows in the corpses, it was apparent they had been taken by surprise. Only elves would be able to ambush a moredhel scout, James was certain, so hopefully that meant Duke Martin and his companions were in the area.

'Do you think the elves will remember us from our last visit to Elvandar, or mistake us for Quegan renegades and start shooting?' Locklear asked.

'Why don't you ask them?' replied a voice from behind a tree to his right, before James could speak.

Locklear and James reined in as a tall man dressed in brown-and-green leather stepped out from behind a tree.

'Martin!' said James. 'I'm pleased to see you, Your Grace.'

Martin, Duke of Crydee, once known as Longbow, stood holding the weapon that had given him his name. 'Well met, James, Locklear. Who is your friend?'

Patrus looked around as a group of elves seemed to materialize from behind the trees. 'Patrus is my name.'

'He's a magician, and he's here to help us shut down the rift machine,' answered Locklear.

'This is Lord Martin, Duke of Crydee.'

Patrus nodded. 'Odd-looking Duke, if you ask me.'

Martin smiled a half-smile very similar to his brother Arutha's. 'Court dress isn't particularly useful when running through the woods, I have found.'

'Well, there is that,' said the old man, scratching his neck. 'We're looking for that machine. You have any idea where it is?'

'I know exactly where it is,' said Martin. 'A large company of moredhel left for the south this morning, and we slipped in behind their lines and came back upriver. I didn't see the machine but I saw enough guards in a small area to be pretty sure where it is. Besides, there's an odd feeling in the air, and it puts me in mind of that Tsurani machine in the Grey Towers back during that war. I'm sending word to Gardan so he can bring up the forces with Arutha when they get here.'

'Too late,' said James. 'Patrus is of the opinion – '

'Patrus doesn't need anyone to tell his opinion for him, boy,' said the old man. To Martin he said, 'Duke, that pointy-eared bastard has probably only managed to get a few companies through with him else they'd be cleaning up the woods with the Kingdom forces already here. He's almost certainly on his way to Sethanon, which is where these boys say he wants to go, so you're probably facing some rear guards. But, if that machine stays open until the Prince gets here, well, there's no telling how many more goblins and trolls and other bloodsuckers you're going to be looking at before you can battle your way down to Sethanon.'

Martin was too concerned about what the old magician said to worry much about his lack of protocol. 'What do you propose?'

'Get us close to that damn machine, then set up a diversion and draw as many off as you can, say fake a major attack then fall back slowly, and if we can slip in, I can disable that machine.'

Martin glanced at James, who shrugged.

Martin said, 'Very well, follow us, but you'll have to come on foot.'

The three riders dismounted and one of the elves took the horses. 'Lead them back to the Knight-Marshal and tell him what you heard here. Tell him also that I expect we could use some relief to the west of the old stone bridge.'

The elf nodded, took the horses and left. Martin glanced at the sky through the branches. 'Good, this will bring us to the bridge at sundown, and you should be able to slip across the river in the dark while we distract them at the bridge.'

'Slip across the river?' said Patrus. 'You mean as in get wet?'

'Yes,' said Martin with a smile. 'I'm afraid you'll have to wade. There's a shallow ford about a mile upriver and I doubt the goblins know of it.'

'Goblins?' asked Locklear.

'We're seeing mostly goblins and Quegan mercenaries. I think most of the moredhel went south with Delekhan or whoever is leading that detachment.'

James was silent for a moment, then said, 'Locklear, how's your Quegan?'

Locklear said, 'Pretty fair. How's yours?'

'Not as good as yours,' said James. 'I didn't grow up in a port town like you did.'

'What's Krondor? Landlocked?' asked the younger squire.

'You don't see many Quegan traders in the sewer, is my point.'

'Oh,' said Locklear. 'Why?'

'Just that I don't think we were likely to run into too many goblins or dark elves who spoke Quegan up at Northwarden, but I'm willing to bet you a good meal we're going to find some of those damn Quegans around here.'

'You're not thinking of doing the "we Quegan mercenaries" thing again, are you?' said Locklear with a look of concern on his face. 'That worked fine when we were lying to trolls, but if there are some real mercenaries here . . . I don't speak it well enough to fool a Quegan.'

'We'll worry about that problem if we come to it,' said James.

Locklear rolled his eyes upward and said, 'Great.'

James said, 'Martin, instead of trying to mount an attack, why don't you chase us into the enemy's arms?'

Martin raised an eyebrow in curiosity. 'Are you sure?'

'No, but let's do it anyway,' said James with a grin.

As the sun set in the west, the sound of men running and shouting in Quegan for help reached the mercenaries at the bridge. Three men, two young and one old, raced for the bridge and in the distance pursuing soldiers were evident in the haze.

James was first to reach the bridge. Out of breath and looking desperate, he pointed and, in Quegan, shouted, 'Attack!'

The second young man said, 'Hold this bridge! We'll get help!'

The mercenary leader turned to the troll who commanded his company and was about to ask what to do when an arrow arched overhead, landing nearby. Ducking behind the scant shelter afforded by the sides of the old stone bridge, he turned his attention to the attacking elves as the three who carried the warning ran for help.

James kept going and looked over his shoulder. Martin and the elves were doing their best to convince those at the bridge that there was a major offensive being launched at them.

James halted and motioned for Locklear to stop, allowing Patrus to catch up. The old man was clearly winded, and James asked, 'You all right?'

Patrus nodded. 'Not quite as fast as I once was. Give me a moment, then let's do it again.' He smiled his evil smile. They paused while Patrus caught his breath, then he nodded and they hurried off.

They were running toward where they thought the rift machine was hidden, in a small depression between two sheltering hills. A group of moredhel ran toward them and Locklear shouted in Quegan, 'The bridge is under attack!'

The leader was a tall moredhel, with a set of shoulders to match Gorath's. He wore a heavy fur jacket which left his massive chest uncovered and he shouted. 'I don't speak your Quegan dog's tongue!'

James paused, and took a deep breath. 'I speak the King's Tongue,' he replied, trying to accent his words. 'The bridge is under attack. The trolls sent us for reinforcements.'

'Attack?' He turned to one of his warriors and sent him toward the bridge. 'I am Moraeulf, son of Delekhan and second-in-command to my father. I command here while Delekhan travels to Sethanon. Whom do you serve?'

Locklear glanced at James, and James said, 'We serve – '

'Tell him about the elves!' shouted Locklear, interrupting James as they had rehearsed.

'Elves?' said Moraeulf. 'What is this of elves?'

'And a tall human with a bow, able to hit a man at a thousand yards, is with them,' said James.

'Longbow!' said Moraeulf. 'It can be only Longbow. His death will bring me honour.'

James said, 'What did he call the other?'

'Calin, I think,' said Locklear.

'The Prince of Elvandar is here!' he shouted, grabbing James by the shirt and easily lifting him off the ground.

'That's the name,' said James, not having to work to look fearful. 'Prince Calin, is what he said.'

'Call my guards! We go to kill the eledhel prince and his human friend! I shall eat their hearts!' shouted Moraeulf, dropping James. 'Where are they?'

'At the bridge to the west,' said James, sitting in the dirt.

The six moredhel ran toward the bridge, and James shouted after, 'I'll send the rest after you!'

Locklear helped him to his feet, and said to James, 'I can't believe that worked.'

'It will only work until that hothead gets tired of chasing Longbow and the elves through the wood. I guess we have maybe half an hour. Let's go.'

James again ran through the trees and when he saw another band of moredhel guards near a clearing, he shouted, 'Moraeulf said to come to the western bridge!'

The leader of the moredhel, an older-looking veteran said, 'We are to guard this machine.' He pointed to the rift machine and James realized he was standing in front of it. In the evening gloom and among the trees he would have walked past it had he not been drawn to the guards.

'Moraeulf said we should guard it,' said Locklear.

The warrior cast a sceptical glance at him, but James said, 'He told us to tell you that we Quegan dogs are fit enough to guard the machine; you must come to hunt eledhel. Prince Calin and Martin Longbow attack the western bridge.'

The effect was instantaneous. The moredhel dashed off to the west.

Standing next to the machine, James said, 'I wonder if Calin knows of the high regard in which his dark cousins hold him?'

'I think he's been around long enough to have some idea,' said Locklear. Glancing at Patrus, he asked, 'Can you stop this thing?'

As he asked the question, a squad of six moredhel appeared through the gate, walking down the ramp. James instantly took on a commanding manner and said, 'We are being attacked in the west! Go support Moraeulf!'

They hurried off in the direction he pointed and Locklear said, 'One of these days you're going to run into a moredhel chieftain who just isn't going to believe you.'

Patrus said, 'Let me look at this thing.' He walked around the machine, a platform only six feet deep and ten feet across. Two men could walk through it comfortably side by side, but no more. 'I see they had to bring a lot of stuff here to build this,' said the old magician. 'That's why it's so small.'

James said, 'The one the Tsurani used in the Riftwar was easily six or eight times as big, from what Arutha told me. They could bring waggons through it.'

'This could handle a small cart, maybe,' said the magician. 'Well, let's see what I can do to turn it off.'

He found several devices carved in the wooden pillar closest to him. They were inset with gems. He ran his hand over them top to bottom. When he got to the one second to the bottom, a spark shot out, knocking him over. As James helped him back to his feet Patrus said, 'Well, I learned something.'

'What?' asked Locklear.

'Not to touch that damn crystal again.'

He walked around the machine and as he returned to where James was, another six moredhel appeared. James again instructed them to find Moraeulf and they did as he instructed.

'Can you do this?' asked Locklear. 'Maybe we can set fire to it.'

'That might do it,' said the old magician. Then his voice lowered, and he added, 'But I doubt it. Magic like this doesn't burn easy. Give me a few minutes, boy, and let me study this thing.'

James and Locklear looked at one another and both

silently echoed the other's thoughts: they might not have a few minutes to give.

Pug took a deep breath and said, 'Owyn, you're going to have to help me again. My powers are coming back, but I can't rely on them right now to take us to Sethanon. You will have to transport us there.'

'Me?' objected Owyn. 'I don't know how.'

'Yes you do,' said Pug. 'Much of what I know, you know. You just don't realize it yet. Now, relax and let me guide you.' He held out a metal orb.

Since having Gamina's mind-touch to link them, Pug was able to reach Owyn and help him focus his magic. 'You have to reach out. Sethanon is out . . . there,' he said softly while Gorath watched.

Pug almost whispered. 'You have to reach out and – '

Suddenly Pug's eyes widened. 'The rift! I can feel it!'

Gorath said, 'Where?'

'Somewhere near Sethanon! That is where Makala and the others must be operating.'

'Let us go there!' said Gorath. 'If Delekhan and his allies are there, that is where I must go.'

Pug nodded and gripped Owyn's shoulders. 'Close your eyes and let me guide you. This is just like stepping through a door.'

Owyn did as he was asked and in his mind's eye he saw the doorway. He felt Gorath's hand rest atop Pug's as mentally he stepped through the door.

And fell through the void.

Then landed unceremoniously on the ground.

Looking up, they saw James turn to Patrus and say, 'How did you do that?'

James held out his hand to Owyn and Pug, and Locklear did the same for Gorath. James said, 'If there was ever a more propitious entrance, I've never heard of it. You are welcome.'

'Thank you, James,' said Pug. He glanced at Patrus and said, 'Who is this?'

'Name's Patrus,' said the old magician walking past Pug and studying the machine. 'One side.'

James said, 'This is Pug.'

'I know,' said Patrus. 'Saw him from a distance once. Came to Timons looking for magicians for his Academy. Hello, Owyn.'

'Hello, Patrus. How did you get here?'

'Long story.'

Another party of moredhel appeared in the rift machine and James shouted, 'Elves are attacking Moraeulf to the west! Go aid him!'

The moredhel were disoriented by the passage and one glanced from James to Gorath. Gorath shouted, 'You heard him! Run!'

The moredhel ran.

'They're coming through at about five-minute intervals,' said James. 'No more than half a dozen at a time.'

'That's still seventy an hour,' said Locklear, 'and Arutha is at least a day away. Martin and some elves are keeping things lively to the west of here, and Gardan is coming from the north, but unless we close this thing down, we're going to have a thousand or more moredhel here by this time tomorrow.'

Gorath asked, 'Where is Delekhan?'

'He and his honour guard are already on their way to Sethanon,' said James, 'according to his son.'

Gorath said, 'We must go there!'

'First we must stop this machine,' said James.

Pug said, 'I can help, but I am without power to do it by myself.'

Owyn asked, 'What do I do?'

Patrus jumped up on the platform and said, 'You stand there and keep a safe distance, boy.'

Pug asked, 'What are you doing?'

'You know what has to be done, magician. This sort of thing can't be done gracefully or without risk.' He smiled at Owyn and said, 'Pay attention to him, boy. He knows a thing or two.' Then he jumped into the gate.

'That's the wrong way!' shouted Pug.

But rather than rebound as Pug expected, Patrus seemed to enter slowly into the gap between the poles of the machine. His staff began to glow brightly and with a wild-eyed determination, he shouted, 'You know what you must do, boy! One of us must do this, and I'm old and near the end, anyway. Do it!'

Pug gripped Owyn hard on the shoulder and said, 'Give him your strength!'

Owyn asked, 'What?'

'He can only close the rift from within! Give him your strength! Make it quick!'

Owyn closed his eyes and let Pug guide him once more. He raised his hand and a flow of energy ran down his arm and out his fingers and struck the old magician as he hung halfway in the gate. Patrus shouted; it was unclear if it was a scream of pain or a yell of triumph. Then the rift blinked out of existence and a roar of wind exploded through the poles, knocking them all to the earth.

James was the first back on his feet, looking around. 'Where'd he go?'

Pug shook his head. 'He's gone.'

Owyn said, 'Gone?'

'He knew what he was doing,' said Pug. 'I once closed a rift from inside. I had the help of Macros the Black, and I barely survived.'

Gorath said, 'He died bravely.'

Owyn let the pain wash over him a moment. Patrus had been his first teacher and while he was a gruff old man with few endearing qualities, the boy respected and admired him. After a long moment, he sighed and stood up.

James said, 'I don't know how long it is before Moraeulf

and the others return. But that wind that just blew through here had to alert someone.'

Pug said, 'We must leave anyway.'

'Which way?' said James. 'I have no desire to try to work our way through the moredhel lines again. I've run out of clever ploys and misdirection.'

Pug said, 'We must go to Sethanon.' He looked at Locklear and James. 'You know why, and soon the others will, too.'

Gorath said, 'If Delekhan is in Sethanon, then so must I be!'

Pug said, 'Form a circle.' They did and joined hands. 'Put your mind at rest, Owyn, and I will again guide you.'

Owyn did as he was bid and Pug led him mentally. It was getting easier for the young magician to do this, and he responded more quickly to Pug's direction.

Suddenly they were somewhere else. They felt a sensation of movement, and displacement. 'We are almost – '

They struck a barrier of pain and even Gorath screamed in agony as they were repulsed from their destination. The powerful moredhel was the last to lose consciousness as he saw the others lying on the cold soil, stunned and twitching in pain. Then he fell into a mindless stupor.

TWENTY

Retribution

The ground was damp.

Pug rolled over on his elbow and saw that James and Locklear were still unconscious. Gorath was awake but disoriented. Owyn sat up, his head held between his hands. 'What happened?' he groaned.

'We hit some sort of barrier.' Pug closed his eyes as he felt the pain in his head dissipate.

James slowly regained consciousness. He sat up, blinked, and finally focused his vision. 'Where are we?' he asked.

Pug stood and looked around. 'We are in the old courtyard at Castle Sethanon.' He pointed toward the burned-out gatehouse of the abandoned keep and said, 'Directly below is our objective.'

'How do we get there?' asked Gorath.

'We probe the boundaries of the barrier and find a place where we can get into the ancient tunnels below this city. They can't have erected it over the entire warren.'

'Why not?' asked Owyn.

'Because even six Tsurani Great Ones could not keep a barrier like that up and intact all day and all night. Makala doesn't know I've escaped from Timirianya. So this barrier was put in place to keep anyone from surprising Makala. Which means his six magicians are taking turns keeping it in place. It must be small for them to do that.'

Locklear said, 'Sounds logical to me.' He stood up and groaned.

'If there's one place that might connect with the ancient

tunnels, it will be the lowest level of the abandoned keep. Let's start looking there.'

'I'll get something to make a torch,' said Gorath.

While they waited, Pug said to Owyn, 'How do you feel?'

'Beaten up, tired, and angry. But otherwise fine. You?'

Pug put his hands together, then separated them by inches. A blue spark leaped from one hand to the next. 'I feel my powers returning, slowly. Perhaps this delay will serve us.' Lowering his voice, he said, 'I don't want the others to know, but if we're facing the Tsurani Great Ones, even at the height of my power, we'd be overmatched. We must trust stealth and surprise. If we can close in on any magician, engage him physically, we can prevent him from casting a spell.

'Additionally, we have another advantage. The idea of being physically attacked is totally alien to the Great Ones, who view themselves as almost godlike in their power. They are so conditioned to having their word obeyed without hesitation that if they attempt to command us rather than cast a spell at us, we gain advantage.'

Owyn said, 'I'm not particularly eager for this confrontation. Some of your knowledge is beginning to manifest in my mind, and I think I can do some things now I couldn't have yesterday, but I'm still uncertain.'

'Then follow my lead.'

Gorath returned with torches. 'I found these bundled in an abandoned storage shed over there.' He also sported a Kingdom crossbow and quiver of bolts. 'I also found these.' He tossed the crossbow to Locklear who caught it and examined it.

'It's dirty and hasn't been oiled in ten years,' said the squire. 'But nothing looks rusty.' He put the head of the bow to the ground and put his foot in the metal stirrup designed to hold the bow in place as it was cocked. Unlike the heavy crossbows that needed to be cranked, this light bow needed only to be drawn. 'And I have little faith in this bow wire.'

But the old bow cocked with a loud click and Locklear loaded a bolt into the groove down its length. 'Stand back. If this wire breaks, someone could get hurt.'

He aimed at a nearby door and pulled the release. The bow shot the bolt with a satisfactory thud into the door.

Locklear looked at the weapon with approval. 'I guess they built this one to last.'

'Do you want to test it again?' asked James.

'No,' said Locklear. 'That might be pushing our luck. If I can get off at least one surprise shot, that might make a difference.'

James nodded.

Pug looked at his small band and said, 'Let's go.'

Pug paused, and said, 'Wait.'

They were in a deep tunnel, barely wide enough for them to move through without turning sideways. Gorath's shoulders rubbed one wall or the other as he walked. They had found it behind a flight of stairs, down at the end of an ancient stone tunnel under the castle.

'What is it?' asked James.

'Here,' he said, pointing to a bare wall. 'If I remember, this is where we should find a doorway down into the lower chamber.'

James pulled out his dagger, Gorath did likewise, and the others stepped back as they attacked what looked at first like a blank wall of earth. Soon both man and moredhel were sweating, and those on either side of them were pulling back the earth they dislodged. Then James's dagger point struck rock.

He cleared away the dirt and said, 'I think this is masonry.'

Owyn said, 'Move away,' and held the torch close to reveal old bricks.

Gorath ignored the heat of the torch and leaned close. 'This looks to be crumbling away.' He pushed hard on a brick and it moved with a protesting grind. 'Stand back,' he said. After they had moved down the tunnel a little way, Gorath put two hands against the bricks and pushed as hard as he could. With a low, grinding rumble, first one, then two, then half a dozen bricks fell away from him.

Gorath managed to keep his balance and pull back just as a section of wall gave way. The tunnel filled with fine dust, which made Locklear and Owyn sneeze.

Gorath didn't hesitate. He grabbed the torch out of Owyn's hand and stepped through the hole. Pug and the others followed. The chamber was vast, empty and the dust of ages lay upon the floor, undisturbed for eons. Pug held his hand up and light sprang from it, illuminating the entire area.

It was no natural cave. The ceiling had been carved and in the walls were reliefs of dragons and creatures in armour who rode them. 'Valheru!' whispered Gorath in awe. 'This was once their place.'

Pug said, 'Before we go any further, I must prepare you for what we are going to encounter, not only the risk of facing the Six, but regarding other issues, as well. Located nearby is an artifact known as the Lifestone.

'This artifact was crafted by the Valheru, as a weapon to be used against the gods during the Chaos Wars. It is far beyond my understanding, and I have been studying it as time permits for nearly nine years. But this I know: it was crafted to be a thing of great destruction. It was this item that was the goal of the false Murmandamus during the Great Uprising ten years ago.'

'False Murmandamus?' asked Gorath, obviously confused.

'He was no true moredhel. He was a Pantathian Serpent Priest whose form was changed by dark magic to gull your people into wasting their lives in his cause. He captured their dying life essence so he could use that power as a key to activate the Lifestone. Had he reached his goal, I fear the results would have been the obliteration of all life on Midkemia. The devastation of Timirianya would seem a garden compared to the barren rock that this would have become.'

Gorath looked murderous. 'So many dead because of the Pantathians!'

Owyn was also confused. 'I don't understand something. How could even a priest or magician of high art activate something that was a weapon against the gods? If the Valheru are gone, isn't the secret of this Lifestone gone with them?'

'No,' said Pug. 'The souls of the Valheru are bound within the stone and it may be that tampering with it will free them. Even if they lack bodies, the energy of their combined minds might be enough to use the Lifestone. We don't know, but it's a risk we cannot allow.'

'So Makala wants to destroy us?' asked Locklear.

'He is not mad enough for that,' said Pug. 'But he is blindly loyal to the Empire and thinks the Kingdom harbours a weapon of destruction that some day may be unleashed upon his nation. He is desperate to discover the secret of that weapon so he may either defend against it or build another for Tsuranuanni so they can treat with the Kingdom from a position of strength.'

'The fool!' spat Gorath. 'What a petty mind he must have.'

'Perhaps petty in his view of the universe,' said Pug, 'but powerful and gifted in magic. At my peak, he could not stand against me, but in my weakened state, I may be overmatched. This is why we must dispose of his six companions and then face him, Owyn and I, together.' He looked at Gorath and Owyn. 'I place a tremendous trust in you two, a renegade moredhel chieftain from the Northlands and the youngest son of an eastern noble. Only the Royal Family and a few who were at the Battle of Sethanon, such as Locklear and James here know the secret of the Lifestone.'

Gorath said, 'I will die before I reveal this to anyone.'

Owyn could only nod.

'Now, follow me.'

Pug led them down the long hallway, obviously once the surface entrance to a vast underground city. 'The cities to the north, Sar-Sargoth and Sar-Isbandia were built by the glamredhel in imitation of this one. This was once called the City of Drakin-Korin.'

'Even in our lore, we know that name,' said Gorath. 'Even among the Valheru he was considered mad.'

'Yet it was he who convinced them to give their essences to the Lifestone.'

The tunnel was massive, and Owyn asked, 'Why is this so big?'

Pug smiled. 'Ever seen a small dragon?'

'No.'

'This is a snug fit for a dragon, and the Valheru rode large ones.'

They came to a pair of massive doors, ancient wood as hard as iron from petrifaction. Hinges the size of a man's body had frozen centuries before. There was enough room for them to walk between the doors, and inside a huge hall they stopped.

Suddenly, Gorath was moving, his sword coming from its scabbard and before Owyn or Pug could mouth a spell, two goblins lay dead in the midst of a vast room.

Pug said, 'This means we are close.'

'It also means Delekhan is near,' said Gorath.

'Makala may be using him,' said Pug, 'but I doubt he would reveal the final secret of the Lifestone to him. None of your moredhel witches could transport him down here. He would have to find a way from the surface.'

Gorath said, 'I doubt this complex has only one entrance.'

Pug said, 'True. Makala could transport by magic once he knew where to go, but the first time he came here, someone had to guide him.'

'Nago,' said Gorath. 'He was in the south for nearly a year

before this madness began. If we could get this Makala to speak, I would wager that it was Nago who showed your Tsurani how to gain entrance to this place.'

Pug said, 'We can speculate on this later. However they met, they decided their purposes were enough in sympathy that they could co-operate in this endeavour.' Pug looked off into the gloom, as if trying to see something and said, 'I think, however, that Makala is using Delekhan as much as the false Murmandamus used your people, Gorath. He's sending your people to die fighting my people, to keep the Prince's army away from this place.'

They started walking again, and suddenly Pug said, 'Wait!' He moved forward, put his hand out and commanded, 'Owyn, feel this.'

Owyn came over next to Pug and put his hand out. He felt energy below his palm, a tingling sensation that grew tangible if he pushed on it. 'Is this the barrier?'

'Yes,' said Pug. 'This is what we struck when I tried to help you transport us into the chamber on the next level down.'

Pug pushed a bit and with his hands moved first right, then left. After a moment he was satisfied. 'This is a sphere, and we must walk around the circumference, until we find those who erected it.'

He moved all the way to the left until they encountered a wall, then back to the right. At the far right extreme, Gorath spied a door a short way back. 'Let's try that,' suggested the dark elf.

They entered a tunnel and moved deeper into the earth.

The magic in the room was so powerful even James and Locklear sensed it, making their skins crawl. 'What is this place?' asked James.

'A treasure trove,' said Pug. 'One of many. Touch nothing here. Some of these items are magic and I can't judge the consequences of their inadvertent activation.'

Owyn said, 'What is that?' He pointed to a large hunting horn with a strange runic symbol on it.

Gorath said, 'The inscription is familiar to me. It is that after which we fashioned our own script. It is Valheru.'

'What does it say?' asked Owyn.

'It is the glyph of the Tyrant of Wind Valley.'

Pug tried to remember which of the Valheru that was, and knew Tomas could tell him. 'This is a place of plunder,' said Gorath. 'Prizes and trophies were gathered here.' He looked down at the dust-covered booty; gold, gems and many items both commonplace and alien.

Owyn reached out and held his hand above the hunting horn. 'Pug, please, examine this.'

Pug gently touched the horn and picked it up. 'It exudes magic,' he said softly.

Then Pug remembered, or the object had the ability to place a memory within him. He dropped the horn as if it had suddenly grown hot. 'Algon-Kokoon! Slain by Ashen-Shugar.' Softly he said, 'Tomas would indeed remember this. It is a hunting horn, which . . .' His eyes widened. He took the horn and put the golden cord which held it around Owyn's shoulder, letting it hang at the young magician's side. 'If it still works, it could tip the balance.' Pug glanced around. 'There are so many things here in Sethanon that I have not had time to study. There is so much I don't know.'

Owyn said, 'But we know that Makala lies somewhere over there, and we must stop him.'

Pug nodded and turned to leave the ancient treasures behind, and they moved down another corridor.

A small chamber glowed with light in the distance and Gorath extinguished the torch. To conserve energy, Pug had stopped using his mystic light. He felt full knowledge was returned to him, and Owyn's abilities had grown far beyond what the young boy who had met Locklear months before had possessed, but they knew they faced seven Great Ones from Kelewan, Makala and his six companions.

They crept down the hall and came to a chamber. Gorath peeked around the corner and pulled his head back. He held up three fingers and pointed to Pug and Owyn. They nodded in understanding. Two magicians rested in the next room, with either a servant or guard. Pug had agonized over how to approach his former brethren. He was almost certain Makala had not told them the full story; even if he had, it was not the truth, but rather Makala's warped vision.

Still, Pug had finally decided, they had lent their talents to events which had resulted in the deaths of thousands: humans, moredhel, troll and goblin, and that could not go unpunished. Pug nodded, pointed to himself and motioned to his left, then pointed to Owyn and motioned to the right. He pointed at Locklear's crossbow and held up three fingers, indicating the third person.

He held up his hand a moment, and when everyone nodded, he made a chopping motion and entered the room.

Owyn and Pug were already incantating their spells when the three figures looked up. Two were wearing black robes and the third was a moredhel warrior.

Locklear raised his crossbow, took a breath, let it out, held it, aimed the bow as he had been taught, then pulled the release.

The bolt flew through the air and took the moredhel in the chest, propelling him across the room. He hit the wall with a sickening sound, and slid down, leaving a crimson smear on the brickwork.

The two Tsurani were immobile, unused to danger and having to cope with surprise. The two spells of immobilization went off within a second of each other, and the two figures moaned in pain as they were engulfed. Gorath had his sword ready and stepped forward to kill both magicians.

It was over in moments.

Pug looked around, held up his hand for silence, and listened for any alarm.

It remained silent for a minute.

He said, 'That leaves four, plus Makala.'

James said, 'This looks like a bedchamber.' He pointed to two pallets on the floor. 'Here they rest, while their brothers maintain the shell around the Lifestone.'

Pug closed his eyes and extended his senses. In the distance he detected a familiar presence. He reached out to it, but was prevented from making contact. 'Not yet,' he whispered.

'Not yet what?' asked James, his face starting to show the fatigue of the past few weeks.

Pug looked at him, then at Locklear and said, 'How have you been caring for yourself since you left Northwarden?'

'Arutha's healing priest gave us powder to drink at night, and we awake refreshed after a few hours.'

Pug said, 'Those work for the short term, but when this is done, you'll need to rest for several days. Be wary. Your senses are dulled, and you are not as quick as you think.'

Locklear looked at James and said, 'If he's telling us we're tired, that's not exactly a surprise.'

James grinned and patted his friend on the cheek, roughly. 'He's telling us not to get overconfident, Locky.'

'Jimmy the Hand, overconfident? Heaven forfend,' replied his friend, dryly.

'Come,' said Pug. 'A spell as powerful as this barrier is no trivial thing. It is much like the barrier erected around me by the Timirianyan god.'

They moved down the corridor and entered a large chamber. Figures moved in the distance, and Pug motioned for his companions to spread out.

Suddenly light shone in the cavern.

Two robed figures advanced and, across the room, a voice spoke. 'We were told to expect you, Milamber.'

'Do not oppose me, Zatapek. Makala has lied to you and

you are hip-deep in the blood of innocents. Stop now before you drown in it.'

'Milamber, Makala is not the only one in our Assembly who believes you to be a false Great One who is more interested in his birth nation than the Good of the Empire. Else why have you hidden this mighty weapon from us?'

The second magician behind Zatapek moved to the side, lowering a staff which he pointed at Pug as if it were a weapon. From behind Pug the sound of a crossbow being fired sounded, and the second magician was spun around, a shower of blood fountaining from his shoulder, as his arm was half-torn from his body. He screamed in agony, and Zatapek reacted.

The Tsurani Great One raised both hands, fingers pointing forward and a cascade of blue energy lashed out, striking Pug hard. He felt every muscle try to contract at the same time, the effect of which was his going rigid and toppling over, striking the stone floor hard as he writhed in silent agony.

Owyn reacted. A large globe of fire erupted from his hand and sped toward Zatapek. But the Tsurani magician was ready, and with a spin of his left hand, he seemed to fashion a shield of energy upon which the fire splashed and flowed to the floor, winking out as it struck the stones.

The only benefit was that he lost control of the magic he had turned on Pug, and the energy vanished, leaving Pug trembling on the stones, still suffering from the pain of Zatapek's magic.

Owyn could not think of what to do next, so he closed his eyes and let his reactions take over. He put out his hand and with a single word sent a column of compressed air hurtling at the Tsurani Great One. For an instant Zatapek couldn't see anything and was preparing for another energy attack, then when he realized what was occurring, he reacted too late. The hammer of the blow slammed him twenty feet across the stones.

Gorath ran toward him, and the last thing the dazed

Tsurani magician saw was the towering figure of the dark elf above him, poised to strike. Then with a single blow, Gorath killed the magician.

Owyn hurried to Pug's side and saw the older magician was still suffering from the lingering effects of the spell cast on him. He shook, and his expression was agony, his teeth were locked and his lips pulled back in a rictus grin.

'What can you do?' asked Locklear.

'I don't know,' said Owyn. He reached out and touched Pug and his fingers exploded in pain. But rather than pull away, he turned his mind to the pain and felt the energy. He moved the energy and turned it to the floor, and suddenly it was gone.

Pug collapsed. Then he took a huge breath, almost a sob, and let it out with a sigh.

James said, 'Pug! Can you understand me?'

Weakly, Pug nodded. Speaking slowly, because the muscles of his face hurt, he said, 'Help me up, please.' Standing with James's arm around his waist, Pug's legs trembled. 'If I move, the effects should pass.'

Owyn said, 'I'm glad. That was more magic in a minute than I've seen in most of my life.'

'You did well. You trusted your instincts. If you continue to do that, the magic you have gained from me will serve us both.' Pug moved away from James and seemed to regain strength with every step. 'That is four of them. At least one of those remaining will be maintaining the barrier spell, if not both. If we can find them, and disable them, the spell that blocks the Lifestone will fall and we can get to Makala.'

He looked around. The magician whom Locklear had shot had died from blood loss. 'Through those doors there is another treasure chamber. It is one which has a blasted wall on the other side, stones destroyed by a mighty struggle years ago. Through that gap in the stones lies the chamber of the Lifestone.'

Gorath said, 'Then we must assume the last two Tsurani guardians and Makala are through those doors.'

'Come,' said Pug. He walked slowly and as he approached the doors, he glanced at Zatapek's companion, a young Great One he had not known. He must have been a trainee when I lived on Kelewan, he thought. Pity. With vacant eyes the dead magician stared up at ancient stones on an alien world. What a waste, thought Pug.

Near the door, Pug motioned for the others to stop. He chanced a glimpse and saw two figures waiting, though he didn't think they had yet seen him. Their attentions were being directed toward two goals, being alert for Pug's approach, and maintaining the barrier behind them. Pug knew that fatigue would be the price of such prolonged duty, but had no illusions as to the time allotted to them.

Delekhan and his own Spellweavers would be somewhere close by, trying to locate this cavern, either to free Murmandamus if they truly believed him here, or to claim the legacy of his powers. Either way, either his arrival or Makala's activating the Lifestone, would prove an abrupt ending to all their efforts.

Pug stepped back and closed his eyes.

I called for you, but was unable to reach your mind, came a familiar voice in Pug's mind.

He looked to Owyn and said, 'The Oracle.'

Owyn nodded.

Pug sent, *We must lower the barrier and free you.*

The Black Robes stole in one night and filled the cavern with a mist that caused my servants to sleep and rendered me weak. Then they bound me with wards that even my powers could not break. It was my inability to know my own future that blinded me to such a possibility. In time I might win free, but so far I am but an echo of what I once was.

Pug considered the might of the Oracle of Aal and was impressed at Makala's preparation. He must not underestimate Makala at any time.

Makala is reckless, and single-minded, but he is not cruel by the standards of his people; had he wished you dead, he would have taken your life already. He is most likely content to have you incapacitated for a time. I think it unlikely you have suffered any permanent injury.

To Owyn Pug said, 'We must quickly eliminate those two.'

Owyn asked, 'Are you able?'

'I must be.' He turned to James, Gorath and Locklear. 'If they are ready for magic, they may not be ready for a physical attack. As soon as we go through the door, hurry after us, but stay to the side in case we are struck.' To Gorath he said, 'What you will see beyond the barrier will amaze and shock you, but do not be alarmed. It is a great dragon, but one unlike any on Midkemia before. She is the Oracle of Aal and must be protected from Delekhan or any other menace while she regains her strength. If I fail to overcome Makala, she is our only hope.'

Gorath nodded. 'I understand.' He looked at Locklear and James. 'These are worthy companions. We shall protect the dragon while you two dispose of the magicians.'

Owyn was about to move to the door, when Pug stopped him. 'There is one spell that may give us the time we need to confront Makala. When the barrier falls, he will know his companions have failed.'

'What is it?'

'If they are as weakened by their labours as I suspect, there is a mild spell that will stun them to senselessness.' Pug pointed. 'If this works, they will continue to hold the barrier in place for a few moments, just long enough for us to cross the chamber to the great rent in the wall between this chamber and the next. We need the time, because to confront Makala, we must shift ourselves in time.'

Owyn nodded. He closed his eyes for a moment, then said, 'I know the spell you speak of.' His eyes widened and he said, 'This seems simple.'

Pug said, 'If we survive this, remind me to tell you how long I researched this particular magic.' He nodded and they stepped through the door. The two Tsurani Great Ones had obviously prepared, for both continued to hold the barrier in place, splitting their energies so that one attempted to defend against Pug, while the other tried to cast a spell of fire at him.

'Look out!' shouted Owyn as he moved out of the way.

The Tsurani may have been prepared for Pug alone, but they didn't expect a second magician. Both Pug and Owyn cast their enchantments, spells which reached out and seized the Tsurani's fatigued minds, stunning them with a blow as effective as if they had been struck by a hammer.

Pug ran. Owyn was a step behind as the two Great Ones stood motionless, rooted and unable to do anything for a moment. Then they slumped to the floor.

As the barrier fell, the room beyond was revealed. Even knowing he was going to see a dragon didn't prepare Owyn for the sight before him. The dragon was immense, easily the largest living creature he had ever seen. Its head rested on the stones, and was the size of a waggon, and its hide was encrusted with gems. Thousands of diamonds covered its body from nose to tail, including its massive wings. But sprinkled throughout were enough rubies, emeralds and sapphires to give the creature a scintillating, rainbow hue that seemed to dance across the surface of her body. Hooded lids covered her eyes, and white teeth the size of sword blades peeked from beneath lips set in a wolfish smile.

Thank you, magicians.

Pug saw a device of Tsurani manufacture that had generated the barrier. In it were tiny bits of crystal. He examined them and said, 'Now we know why Makala was so desperate to have that ruby.' He pointed to the machine. 'Those stones of unusual property you mentioned, Owyn. They were used to power this device, and the Six were responsible for keeping it functioning. I knew there was no simple magic that could disable the Oracle.' To the dragon, Pug said, 'Gorath and the others will stand guard while Owyn and I tend to Makala.'

You must hurry. He acts rashly.

Pug and Owyn ran toward the gap in the wall, smashed out by the struggle that a dragon had undertaken with one of the most terrible creatures known, a Dreadlord, during the height of the battle that decided the Great Uprising and ended Murmandamus's threat to the Kingdom.

Pug faltered. He had expected to see an empty room in which he would have to shift Owyn and himself a moment in time to bring them into phase with the Lifestone. Instead he said, 'Gods! Makala has brought the Lifestone here!'

The Tsurani Great One stood before a large emerald-coloured stone as high as a man's waist. From the top protruded a golden sword with a white hilt that looked like ivory. On the hilt was embossed a golden dragon.

Then the Black Robe turned and said, 'Milamber, I am impressed. I didn't think even you could win past all my defences. I hope Zatapek and the others did not suffer unduly in trying to stop you?'

Pug's anger was barely held in check. 'They died like loyal Tsurani, with honour and obedience, and completely ignorant of your murderous duplicity.'

'Do not speak to me of duplicity, Milamber! You swore an oath to serve the Empire, yet you hide the existence of this terrible lie from us!' Makala moved a step toward Pug and Owyn and shouted, 'Ten years ago you engaged in a battle to bar the Enemy from our worlds, or so you told the Assembly. Hochopepa and Shimone bore witness to that lie. Sons of great Tsurani houses died on this alien world to aid in that great cause. Yet, you denied us any explanation as to why this city was important, why we had to spend Tsurani blood here.'

He lowered his voice. 'When I came to your world you deflected my questions and were evasive, and when at last

I undertook to discover reasons myself, I discovered this place, with its traps and magic wards, and that great creature in the next hall. All here to keep me and anyone else not in your service from that!' His finger shot out, pointing to the Lifestone.

'You did not bar this world to the Enemy! You trapped it in that stone, and harbour it against the day you feel the need to unleash it against your foes, perhaps against the Empire of Tsuranuanni!'

'You can't believe that,' said Pug.

'Not only do I believe it; I intend to make sure that day never comes. I have almost unlocked this thing and when I fully understand its nature, I will take it to the Holy City and there it will wait until the Empire needs it for her defence.'

Owyn said, 'Pug, he's mad.'

Makala said, 'Boy, this conversation is not for children.' He made a dismissive motion with his hand and Owyn was flung backward, as if he had been struck. At the last instant he recognized the spell as a variant of the same one he had used to fell Zatapek and shielded himself from serious harm. But he still landed hard on the stones and had the wind knocked out of him for a moment.

Pug turned and said, 'You're a murderous dog, Makala. I welcomed you into my household and you betrayed me and my trust. You treated my daughter as a pawn in a game and put her life at terrible risk. For that act alone you've earned death. But thousands have died for you to reach this point.'

'All the more reason it's vital I succeed, Milamber. Else they died in vain. When this artifact is safe in the Imperial Palace, they will have died for the good of the Empire.'

Pug gathered his depleted power to him, knowing he was facing one who was among the most gifted of the Assembly.

Makala stepped back. 'I will not face you, Milamber. I was amongst those who was in attendance when you single-handedly destroyed the great arena in Kentosani. I have no delusions of being your equal, even in your weakened state.' He turned slightly and made a signalling with his hand.

From out of the shadows two figures appeared, large menacing grey figures with massive wings. Makala said, 'One useful thing I gained when I discovered the world of Timirianya, was a staff belonging to an ancient priest named Rlynn Skrr. It allows me to command these creatures.' He said to the two elemental creatures, 'Kill them.'

Pug turned to Owyn and said, 'I can fight one, you the other, but we can't fight them *and* Makala. Blow the horn!'

Owyn didn't hesitate and raised the hunting horn to his lips. He blew and a long, plaintive note hung in the air, sounding as much like a dog's howl as a hunting horn.

A chilling wind struck the hall, nothing natural, a thing of ancient magic. Suddenly next to Owyn stood a pair of hunting hounds, massive in size with slavering jaws and fangs the size of daggers. Their eyes were red, and around their necks they wore studded collars of iron spikes. They stood waiting.

'Command them, Owyn!' shouted Pug.

'What do I do?'

Pug turned to face Owyn, and in his eyes the young magician saw anger and hate. 'Makala!' he shouted.

Owyn stood and pointed at the suddenly-unsettled Tsurani magician. 'Attack!' he commanded.

The hounds leaped forward. Pug turned as the first of the wind elementals neared him, and reaching deep within, employed the spell he had used on them before on Timirianya. As before, the creature was engulfed in spinning coloured beads of energy, and stood rooted, wailing a ghostly cry.

Owyn cast his spell at the other, and it, too, stood rooted.

Then they turned their attention to Makala. The Tsurani Great One had erected a protective shield against the great beasts which stalked him and prodded against it, trying to find a way around it. He retreated, and as they closed on him he was prevented from employing any magic against Pug or Owyn.

Pug moved around the Lifestone and took a moment to glance at it, seeing if it had been endangered in any perceivable way. He said a momentary prayer of thanks; apparently Makala hadn't yet begun to interact physically with the gem.

Pug then turned to Makala who sought to avoid the lunges of the hounds. They couldn't reach him, but their attacks were unsettling.

Pug came to stand beside one of the huge dogs and shouted, 'Makala, you have betrayed me, my family, and your own brotherhood in this mindless adherence to a blind Tsurani credo! You did not even bother to determine what was "the good of the Empire". Had you even begun to understand, you would know that what you propose to do places the Empire in the greatest risk it has known since the Enemy drove the Nations across the Golden Bridge. Thousands have died for your arrogance and vanity. For all of this, you are condemned to death.'

With a wave of his hand, he summoned a spell and with it he peeled back the protective enchantment Makala had raised. The older magician realized at the last instant what Pug had done, and screamed, 'No!' Then the hounds leaped on him and began tearing him to shreds.

He died quickly. The dogs continued tearing at his corpse, rending it to pieces they scattered around the hall.

Owyn approached as the two wind elementals faded from view and said, 'He deserved no less.'

'Call off the hounds,' said Pug.

Owyn shouted, 'Stop!' and the hounds stopped. He turned to Pug and said, 'What do I do with them?'

Pug shook his head. 'I think you just need to tell them to go back where they came from.'

Owyn turned and did so, and the two hounds vanished from sight. He took off the horn and put it down. 'This is a terrible power to have.'

Pug put his hand on Owyn's shoulder. 'All power is terrible if not used wisely.' He glanced at the mangled corpse and said, 'That was once a man of great power and position. He abused both. Never forget that.'

Owyn said, 'I never will. I don't think I'm cut out for a magician's life.'

Pug actually managed to laugh. 'Cut out or not, I don't think you can avoid it. You're a young man of great power, Owyn.'

'Me? I'm just a youngster who learned some things from Patrus and from you.'

'More,' said Pug. He put his hand on Owyn's shoulder and said, 'When we linked minds you were given much of my knowledge. You will find that some of it will lie dormant for years, but other parts will come to you unbidden. Whatever you choose to do with your life when you return to your father's court, you are going to be one of the more gifted practitioners of magic in the world.'

Owyn said, 'That will take some getting used to. I – '

Further conversation was interrupted by the sound of swords clashing and shouts from the next chamber.

Magician, came the dragon's thoughts, *I cannot stop them. I am still too weak.*

Pug turned toward the gap between the chambers and saw someone hurrying through it. For a moment he thought it was Gorath, but too late he recognized it was a different moredhel.

This one was carrying a staff which he levelled at Pug and Owyn. A blast of energy smashed them both across the room. Pug hit the wall hard enough that lights danced before his eyes, and Owyn again had the wind knocked from him.

He saw the moredhel Spellweaver struck from behind by someone, and saw Locklear stumble into the chamber, then turning to barely avoid the sword blow of another moredhel, a warrior who had vainly tried to prevent Locklear from reaching the Spellweaver.

Suddenly the room filled with combatants. Locklear fought a moredhel warrior, while James attempted to keep close to another Spellweaver, who tried to fend him off with a staff like the one used to hurt Pug and Owyn.

Dazed, Owyn tried to concentrate and help, but he couldn't focus. He went to where Pug still lay and helped him to his feet, saying, 'I'm getting very tired of that. My back is killing me.'

Pug shook his head and said, 'What?'

Gorath was fighting Delekhan. The moredhel chieftain wore the black helm Pug had seen on Murmandamus and gaudy black armour with gems on the breastplate.

Gorath lost his footing and stumbled and Delekhan struck him hard across the face with his free hand, knocking the chieftain of the Ardanien backwards. Gorath went sprawling across the floor.

Delekhan saw the mangled body of Makala and actually smirked in satisfaction. Then he saw the Lifestone.

Pug realized that he could not get there before the self-styled moredhel ruler. Makala had been dangerous because of what he knew; Delekhan was dangerous because of what he didn't know. He saw the golden sword and his eyes widened. 'Valheru!' he cried. 'It's a sword for a king!'

He lunged for it, only to have Gorath leap on his back, closing his arm around Delekhan's throat.

Delekhan's hand grasped the hilt of the sword and suddenly a thrumming sound filled the cavern. Delekhan's eyes widened and he began to gurgle, but not from the choking Gorath was inflicting on him. Rather, a great power was trying to manifest itself within the moredhel.

The sword began to rise, and Gorath abandoned his attempt to kill Delekhan and instead gripped the hilt and tried to push the sword back into the stone.

Pug shook his head, and saw that Locklear and James still struggled with their opponents. Owyn asked, 'What do we do?'

'Gorath! Stand clear,' shouted Pug.

'I cannot,' said Gorath. 'If I do, he will pull the sword free.'

Both moredhel struggled, the muscles and cords on their arms and shoulders bulging from effort. Delekhan's eyes widened to impossible size, as if they were about to burst from his skull, his face was flushed and perspiration poured off his skin. An alien cast came over his features and it looked as if another person was now wearing his face.

'He's transforming!' shouted Pug.

Owyn said, 'We must stop him!'

'Do not touch him!' shouted Pug over the increasingly loud noise.

'We must help him!'

'We cannot,' said Pug. 'You must help me. We must destroy them both.'

Owyn said, 'I can't.'

Gorath shouted, 'You must! Ancient powers are seeking to take my life! Save my people, Owyn. Save me.'

Owyn nodded and with tears welling in his eyes, he quickly moved his hands in a complex pattern above his head. Pug duplicated the movements, and as one they pointed to the two moredhel locked in a death-struggle before the Lifestone.

A blast of heat erupted from the magicians' hands, and a white-hot light struck both moredhel. For a moment they stood bathed in eye-searing brilliance, so bright that James's opponent turned his head away, and James managed to step close and drive his dagger deep into the magician's chest. James turned toward the light, and was forced to look away from the brilliance. Shielding his eyes with one hand, he moved and struck the moredhel warrior Locklear was facing in the back of the head, causing him to falter, and Locklear finished him off.

A low moan came from the two figures in the light, then they faded from view. A moment later, the light flickered out.

Again, the stone was untouched and the sword remained in place.

Silence descended on the chamber and the four men in the room fell in place, exhaustion threatening to overwhelm them.

Owyn wept and Pug said, 'I think I understand.'

Locklear said, 'What happened? I couldn't see.'

James looked around the chamber and said, 'Gorath?'

Pug said, 'He saved us all.'

James nodded, his expression bitter. 'I will never think of the moredhel in the same fashion.'

Locklear sat and said, 'He was a difficult companion at times, but he was . . . a friend.'

Pug was too numb to move. 'I think I'll sleep a week,' he said.

James said, 'Catch your breath, m'lord Duke of Stardock, for we have work left to do.'

Pug said, 'Work?'

With an evil grin James said, 'Have you forgotten that stone has to be shifted back where it belongs? And there's a moredhel army still in the Dimwood? And Delekhan's advance guards are all around us?'

Pug said, 'I'm trying.'

Owyn said, 'If they show up now, I'm dead. I can't lift a finger.'

Locklear said, 'Well, if we're to survive all this, I'd rather not die because I'm too tired to defend myself. Can either of you magicians think of something?'

Pug said, 'I can. Help me up.'

James pulled him to his feet and asked, 'What are you going to do?'

Pug said, 'With whatever strength we have left, my friends, we are going to put on a show.'

Locklear stopped and blinked in confusion. 'My mind is going. For a moment I thought I heard you say we were going to put on a show.'

'That's what I said,' Pug said. 'Come with me.'

Three fatigued, confused men exchanged glances with one another, then followed the strange short man in the black robe.

Moraeulf was furious. He had been in a running fight with Prince Calin and Longbow for two days, but had yet to close with them. In the mountains, the moredhel had the advantage, but here in the heavy woods, the eledhel and their demon human friends had the upper hand.

The only good thing in this had been the course of the fight, which had taken them to the edge of the city of Sethanon. Moraeulf was waiting for his father's orders, and word had reached them that somehow the rift machine had been disabled. Heads would sit on pikes over that, and Moraeulf was determined that his wouldn't be among them.

'Master, runners come.'

He expected to see his own scouts who had been trailing the eledhel, but instead two of his father's honour guards approached, dirty, tired and obviously near panic. 'What is it?' he demanded.

'Disaster! On the walls of Sethanon!'

'Tell me!' shouted Moraeulf.

'Three days ago we found our way into the city and our master left us near the rear gate of the castle. For most of a day he was gone. Then came a great sound from deep within the earth, and then we saw something terrible on the battlement of the castle.'

Moraeulf grabbed one of the guards by the shirtfront and demanded, 'Tell me what you saw!'

The other said, 'On the battlement we saw your father, and with him was Murmandamus. I know it was he, for he was without shirt, and the dragon mark was on him. He was gaunt, as if he had been starved, and pale, as if kept underground, but it was he. There could be no doubt. He shouted and we could hear his voice, carried to us by magic as we had heard him ten years ago, lord; and it was his voice.'

'Aye,' said the other guard. 'It was Murmandamus. And between him and your father stood the human prince, Arutha, in their thrall. Murmandamus said he would at last fulfil the prophecy and end the life of the Lord of the West, but as he drew back his blade – '

'What?' shouted Moraeulf, striking the warrior, knocking him to the ground. 'On your life tell me,' he demanded of the other one.

'My lord, from behind rose a great dragon, a creature the like of which no living being has seen. It was afire with light and covered in rainbows and on its back rode a magician in black. He cried out that Murmandamus was a false prophet and the prophecy was also false, and then the dragon unleashed a blast of fire so hot we could feel the heat of it on the ground below.

'Lord, your father and Murmandamus were withered before our eyes, turned to ash and blown by the winds, while the Lord of the West, the human, Prince Arutha, stood unharmed!'

Moraeulf howled his rage and struck the man. 'Damn all magicians and prophets!'

There were half a dozen warriors of his own with the two from his father's guard. 'Pass word,' he ordered them, 'we return to the north. This madness is over!'

The eight moredhel hurried off to spread the order.

Moraeulf turned to find his way northward to his main camp. He was only a few yards along the trail when a shape stepped out of the gloom and asked, 'My lord?'

'What?' demanded Moraeulf. Too late he realized he knew the person who closed on him, and recognition came with pain, as Narab drove his dagger into the son of his enemy.

Moraeulf sank to his knees, his mouth open in disbelief, and he fell to the earth.

From beyond the path, a voice said, 'We have done our part.'

Narab turned. 'I will do mine.'

Martin Longbow and his elves appeared and Narab said, 'My family is avenged and I will take our people home.'

'We will not trouble any of you as long as you're moving north,' said the Duke of Crydee. 'Never again return south.'

Narab said, 'Liallan and her Snow Leopards and my own clan are now the power in the north. As long as we rule, we shall keep to our side of the mountain.' Then he pointed a finger at Martin and the elves. 'And you would do well to stay on your own side also.'

He turned and vanished, and Martin said to the elves, 'Let us go to Sethanon and find out the mystery behind the wonder we just heard. I would like to find Pug and discover how Murmandamus came back from the grave long enough to be killed again.'

The elf to whom he spoke nodded, his expression conveying his own curiosity.

Martin started walking south. 'At least when my brother gets here, he'll find his Kingdom still intact. I think that will please him.'

Martin Longbow, brother to princes and kings, shouldered the weapon which had given him his name and hummed a nameless tune. He didn't know the details yet, but he knew they had won, and that, for the time being, a future existed for his wife and daughter. That was cause enough to hum a tune; the details would come later.

Dedication

Arutha raised his cup.

'Gorath!' he said.

The others in the command tent raised their cups, said 'Gorath!' and drank to his memory.

Pug had related the final hours of the struggle and how he and Owyn had fashioned the illusion to convince the moredhel that Murmandamus was at last dead. Over the meal he had explained about Gorath's self-sacrifice, his nobility.

Arutha reflected a moment on what he had been told, then said, 'I find it very strange to consider any Brother of the Dark Path noble, but there is no other word for his deed. Even when I sent him off with Owyn and James, I harboured lingering doubts. I could not rid myself of the notion it was but another convoluted plan of the Pantathians. I was wrong.'

'You are supposed to be suspicious,' said Martin. 'It is part of your duty to your Kingdom.'

Arutha sipped his wine and nodded at his brother's remark. 'Perhaps, but I am certainly never going to regard the Brotherhood of the Dark Path quite the same again.'

Owyn said, 'If I may, Your Highness?'

Arutha gave the young magician permission to speak.

'I travelled with Gorath for a few weeks in his homeland, and it's so very different than I could have imagined.' He told of the human communities living side by side with the moredhel, and while the humans would be labelled renegades, it showed that it was possible for peace to exist between the races. 'They have a fierce way, it seemed to me, yet it was a way that is not so alien that we can't appreciate it. I met Delekhan's and Gorath's wives; Delekhan's is a powerful leader in her own right, and she is beautiful and fierce herself, and strange. Yet there was only ambition to save her people and she helped us.' Owyn sighed. 'When I first met Gorath, he said we would never understand his people or their ways. Maybe he was right, but I can accept them.'

Arutha said, 'Would that all of them were like him. Life in our Kingdom would be far calmer in the north.' The Prince continued, 'Those of us who have survived yet another attempt by dark forces to destroy us must again rededicate ourselves to protecting our nation. Otherwise those who have died will have paid a great price in vain. Gorath will be remembered, not as a traitor to his people, but as one of our nation's heroes as well as one of his own. He started out serving only the moredhel. He died to save us all.'

Owyn said, 'I just wish he could have lived the rest of his life in Elvandar.'

Martin said, 'That is something for anyone to desire. It is a good place to end one's days. But Aglaranna was right; he wasn't fully returned, and it was his hatred for Delekhan that prevented him from being one with the eledhel.'

Arutha said to Pug, 'I wish I could have seen that little drama you concocted to deceive the moredhel.'

'That was Owyn's doing as much as mine. I have never been an adept at illusion, but he had learned some of those skills at Stardock. He fashioned the images of Murmandamus – from my description of him, and he had seen Arutha and Delekhan personally – and the Oracle had revived enough to rise up with me on her neck and spout an impressive flame. We just hoped it would be effective enough.'

Martin said, 'Certainly it was that. I was a dozen feet away when I heard two of Delekhan's guards tell Moraeulf of his father's death and the end of Murmandamus. They were believers. Even Narab likely believes the tale. Even if he doesn't, it serves his purposes to let others believe. I think we are done with the moredhel seeking Sethanon again.' Martin put aside his wine and said, 'I must leave. I have a long journey back to Elvandar before me, and then on to Crydee. My elven friends and I will depart at first light tomorrow, so I am to bed.'

Arutha stood and embraced his brother. 'We see you too rarely,' he said.

'Come to Crydee. You and Anita. Bring Borric, Erland and Eliena. Spend a month.'

'Two weeks, perhaps, and not until after the baby's here.'

'Another baby!' said Martin with a grin. 'When were you getting around to telling me?'

Arutha smiled and said, 'I expect there's a message with the royal seal of Krondor waiting for you in Crydee, where you'd have already read it if you were acting the part of a proper duke and not running around in the woods like your elf friends.'

'If I hadn't been running around, those two – ' he pointed to James and Locklear, who had unceremoniously fallen asleep on cushions in the corner of Arutha's pavilion ' – would never have made it to destroy that machine and these woods would be full of Delekhan's warriors.' He lowered his voice. 'And I would not have been able to find Narab and help him get to Moraeulf.' Turning to happier matters he asked, 'Have you picked out names?'

Arutha nodded. 'If a girl, we shall name her Alicia, after Anita's mother. If a boy, Nicholas, after great-grandfather.'

Martin said, 'I look forward to the news of the birth.'

Arutha hugged his brother again, and replied, 'I know. Be well and give my love to Briana.'

Martin departed and Arutha looked at his sleeping squires. 'I think this constitutes lèse majesté. What do you think, Pug?'

'I think the magic herbs your healer gave them wore off. I think your threatening the hangman's noose wouldn't waken them.'

'I'm glad to hear that,' said Arutha. 'They are forgiven.' Looking at Owyn, Arutha asked, 'What shall we do with you?'

Owyn said, 'Highness, I am overdue at home, and really must return to face my father. Not that I can imagine his wrath being any greater, but the longer I tarry the worse will be my punishment for disobeying him.'

Arutha rubbed his chin and said, 'Well, then, perhaps a good horse and some gold for a better inn or two along the way. And I think I'll send a personal note to your father proclaiming my personal indebtedness to the son of the Baron of Timons for his great service to the Crown. I will recommend that if your father can't find a place for you in his service, he consider commending you to the King, so that you might serve the Crown in Rillanon. I'll also send a note to my brother telling him of your service. If you really do wish to, I'm sure he'll find service for a bright lad such as you.'

Owyn smiled. Whatever anger his father had felt at Owyn's decision to disobey and run off to Stardock would evaporate before a personal letter of commendation from the Prince of Krondor. Not to mention a commendation to the King. His homecoming had just become a great deal more attractive. 'I thank the Prince.'

Pug said, 'We need to speak some more about things at Sethanon, Arutha, and about what we must do to ensure nothing like this happens again.' He fought back a yawn. 'But right now we need our sleep, too.'

The Prince inclined his head. 'Then you are excused, my friend, and we'll speak again in the morning. Good night.'

They bid Arutha good night and left the Prince's pavilion. Pug walked with Owyn to the tent Arutha had set aside for their use. 'What will you do after you return home?' asked Pug.

Owyn said, 'I'm not sure. I know that my life will never be the same. I've seen too much and . . . it changes you.'

Pug tapped Owyn's head with his forefinger. 'And you have too much up there to let it lie idle. Come back to Stardock. Make sure we see no more mad wizards like Makala.'

'I don't know,' said Owyn. 'I think I would like to know more about these powers of mine, but I also think my father will have much to say about my future.'

'Such is the burden of nobility,' said Pug. 'But you have time to ponder those choices, and you certainly have a great deal more to think about than before.'

'No doubt,' said Owyn, as they entered their tent. 'Truth to tell, one of the reasons I left Stardock was because of all the politics. Your two Keshian students, Korsh and Watoom, they're gathering followers and I can see some very nasty business ahead if you don't break up those factions.'

'As do I, but I'm unsure as yet what to do about it,' Pug admitted.

Pug sat upon his mat and Owyn started to close the flaps. For a moment Owyn paused and looked out at the calm woods around the camp. In the distance he could hear the soldiers of the Kingdom around their fires, and above the trees the stars shone brightly.

He wondered if somewhere out there Gorath was with the Mothers and the Fathers, or in the Blessed Isles.

Wherever you are, Owyn thought as he tied the tent flap closed, *you will never be forgotten*. Then he added, *my friend*.

He turned to his own mat and lay down. Despite the unanswered questions and the countless possibilities still before him, Owyn fell quickly asleep.

Pug looked at the young magician and remembered when he had been that age, wrestling with the great powers Owyn didn't even suspect he now possessed, and wondered which choices Owyn would make.

But whatever those choices, Owyn would make them, and Pug lay down relieved to know that his home and family were again safe. He basked in the knowledge that Gamina was home and that he would soon join his family at Stardock. With that thought in his mind, Pug drifted off to sleep. And it was a good, long, restful sleep.

AUTHOR'S AFTERWORD

The phone rang.

The voice on the other end of the line belonged to my agent, Jonathan Matson. He said there was a fellow named John Cutter, who wanted to speak to me about a game deal. I told Jonathan to give him my phone number at home and forgot about it.

A while later the phone rang again and a pleasant voice at the other end identified himself as John Cutter, a game producer for Dynamix, Inc. a company in Eugene, Oregon.

John had produced the second game in the successful *Might & Magic* series while at New World Computing, Inc. and was itching to do the same for Dynamix, a company known primarily for flight simulations, particularly the popular *Red Baron*. He had been told by one of the firm's founders, Jeff Tunell, that I was a good fantasy author and thought maybe I could write them a game.

I explained to John that he couldn't afford me, but then introduced him to the idea of licensing, and from that point forward, we were on the same wavelength.

That's how *Betrayal at Krondor* started.

Fantasy role-playing games and books both use stories, but in different ways. My experience prior to working with John and his crew at Dynamix was limited to my own involvement with the creation of Midkemia, the fantasy world in which my work resides, and with playing other people's computer role-playing games.

Neal Hallford and John Cutter wrote the game. I got to

review things, but they wrote it. I talked with them about story, gave them ideas, listened to their ideas, and the game took form. But even I had no idea what it would look like, or play like, until it was finished. I got the script, but John had simply printed everything and sent it to me, without any idea of how it hung together. The first time I jumped from the opening narrative to the initial dialogue in the first sub-quest, I was lost. And that wasn't the last time.

When I finally got a look at the finished game, it was at the Drake Hotel in Chicago before my first press interview on the game at the Consumer Electronics Show in 1993. It was a revelation. It was my world, but it wasn't. These were my characters, but they weren't. They came alive and ran around and fought and died and started over and fought again. When it came time to give the interview, I didn't want to stop playing.

The rest, as they say, is history. *Betrayal at Krondor* won awards, sat atop the *Entertainment Weekly* CD game bestseller list for six months, and is considered by many to be the best computer fantasy role-playing game ever created. And most of the credit goes to John, Neal and the team at Dynamix.

When I approached turning *Betrayal* into this novel, I was faced with many decisions, revolving around story elements that make for a really good game, yet are either inconsistent with the literary Midkemia, or are just too silly to believe. 'The Quest for Ale' and 'Find The Lost Minstrel', to name two sub-quests in the game, were clearly going to totally destroy the tension of the story.

That being the case, I decided that rather than attempt to 'novelize' the game, I would take the core story of the game and tell it in novel form. So that's what I did; I took Neal and John's story for the game, *Betrayal at Krondor*, and started churning it around in my head, deciding scene by scene what would go, stay, be changed or introduced. The book you hold in your hand is the central story of the game, without most of the sub-quests and side trips, and without a great deal of what makes a game a game. But the story of Owyn and Gorath, James and Locklear is at the heart the same one.

So for those of you who have played the game, *Betrayal at Krondor*, this novel, *Krondor: The Betrayal*, will be very familiar, but will also contain a few surprises. For those who have never seen the game, just consider this another missing chapter in the ongoing history of the world of Midkemia and the City of Krondor.

Raymond E. Feist
Rancho Santa Fe, CA
March 11, 1998

Talon of the Silver Hawk

Conclave of Shadows: Book One

Raymond E. Feist

Four days and four nights Kieli has waited upon the remote mountain peak of Shatana Higo for the gods to grant him his manhood name.

Exhausted and despairing, he is woken by the sharp claws of a rare silver hawk piercing his arm, though later he is not sure if it ever happened.

Devastation greets Kieli on his return home. His village is being burned, his people slaughtered. Although it means certain death, Kieli throws himself into the battle . . . and survives.

A distant voice echoes in his mind: *Rise up and be a talon for your people . . .*

Now he is Talon of the Silver Hawk, and he must avenge the murder of his people, at whatever cost.

'Feist writes fantasy of epic scope, fast-moving action and vivid imagination'
Washington Post

ISBN: 0-00-716185-9